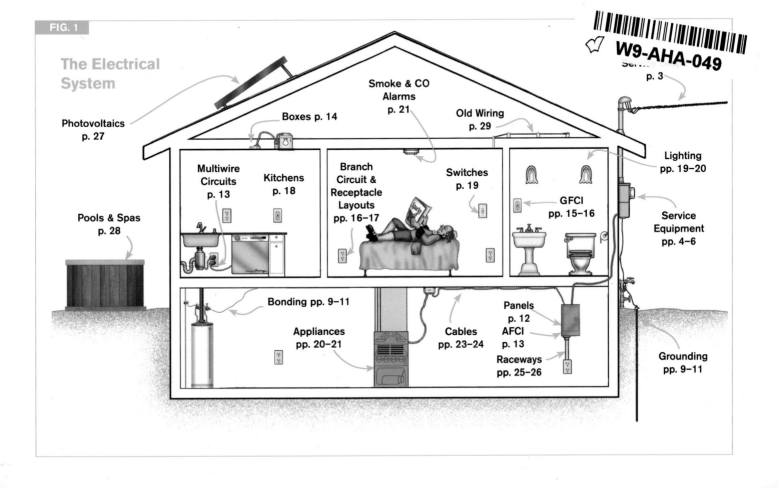

FIG. 1

The Electrical System

Photovoltaics p. 27

Boxes p. 14

Smoke & CO Alarms p. 21

Old Wiring p. 29

p. 3

Lighting pp. 19–20

Multiwire Circuits p. 13

Kitchens p. 18

Branch Circuit & Receptacle Layouts pp. 16–17

Switches p. 19

GFCI pp. 15–16

Service Equipment pp. 4–6

Pools & Spas p. 28

Bonding pp. 9–11

Appliances pp. 20–21

Cables pp. 23–24

Panels p. 12

AFCI p. 13

Raceways pp. 25–26

Grounding pp. 9–11

GLOSSARY OF ELECTRICAL TERMS

Accessible, as applied to wiring methods: Not permanently concealed or enclosed by building construction.

Accessible, as appplied to equipment: Capable of being removed or exposed without damaging the building finish or structure. A piece of equipment can be considered accessible even if tools must be used or other equipment must be removed to gain access to it.

Accessible, readily: Capable of being reached quickly for operation or inspection without the necessity of using tools to remove covers, resorting to ladders, or removing other obstacles.

Alternating current (AC): Current that flows in one direction and then in the other in regular cycles, referred to as frequency or Hertz.

Apparent power: see *power*.

Approved: Acceptable to the AHJ. The AHJ will usually approve materials that are listed and labeled.

Arc-fault: An electric current propagated through air.

  • **AFCI, Arc-Fault Circuit Interrupter:** A device intended to provide protection from the effects of arc faults by recognizing characteristics unique to arcing and by functioning to de-energize the circuit when an arc fault is detected.

  • **AFCI, branch/feeder type:** A "first generation" AFCI device capable of interrupting parallel arcing faults. They do not meet the present code standard.

  • **AFCI, combination type:** An AFCI meeting the standard for interrupting both series and parallel arcs.

Authority Having Jurisdiction: (AHJ) The building official or persons authorized to act on his or her behalf.

Bonded, bonding: Connected to establish continuity and conductivity.

Branch circuit: The circuit conductors between the final OCPD (breaker or fuse) protecting the circuit and the outlet or outlets.

  • **Branch circuit, general purpose:** Branch circuit that supplies two or more receptacles or outlets for lighting and appliances.

  • **Branch circuit, individual:** Branch circuit supplying only 1 piece of equipment.

  • **Branch circuit, multiwire, residential:** Branch circuit consisting of 2 hot conductors having 240V potential between them and a grounded neutral having 120V potential to each hot conductor F17.

  • **Branch circuit, small appliance:** Branch circuit supplying portable household appliances in kitchens and related rooms and that has no permanently installed equipment connected to it (see **p. 18** for exceptions).

Clothes closet: A non-habitable room or space intended primarily for storage of garments & apparel F37.

Controller: A device to directly open and close power to a load.

Derating: A reduction in the allowable ampacity of conductors because of ambient temperatures >86°F, or more than three current-carrying conductors in the same raceway, or for cables without spacing between them.

Device: A piece of equipment that carries or controls electrical energy as its primary function, such as a switch, receptacle, or circuit breaker.

Equipment: A general term including materials, fittings, devices, appliances, luminaires (fixtures), apparatus, machinery, and the like used as a part of, or in connection with, an electrical installation.

Equipment Grounding Conductor (EGC): A wire or conductive path that limits voltage on metal surfaces and provides a path for fault currents F16.

Flexibility after installation: Anticipated movement after initial installation, such as that caused by motor vibration or equipment repositioning.

Feeders: Conductors supplying panelboards other than service panels.

Gooseneck: A curve at the top of a service entrance cable designed to prevent water from entering the open end of the cable.

Ground: The earth.

Grounded conductor: A current-carrying conductor that is intentionally connected to earth (a neutral).

Grounding Electrode Conductor (GEC): A conductor used to connect the service neutral or the equipment to a grounding electrode or to a point on the grounding electrode system F6.

Ground fault: An unintentional connection of a current-carrying conductor to equipment, earth, or conductors that are not normally intended to carry current.

- **GFCI, Ground-Fault Circuit Interrupter:** A device to protect against shock hazards by interrupting current when an imbalance of 6 milliamps or more is detected.

- **GFPE, Ground-Fault Protected Equipment:** A device to protect equipment from ground faults and allowing higher levels of leakage current than a GFCI.

Hertz: A measure of the frequency of AC. In North America, the standard frequency is 60 Hertz.

Individual branch circuit: see *branch circuit, individual*

In sight: see *within sight*.

Interrupting Rating: The highest current a breaker or fuse can interrupt without sustaining damage.

Lighting outlet: An outlet intended for the direct connection of a lampholder or a luminaire.

Load: The demand on an electrical circuit measured in amps or watts.

Location, damp: An area protected from the weather, yet subject to moderate degrees of moisture, such as a covered porch.

Location, dry: A location not normally subject to dampness or wetness.

Location, wet: All areas subject to direct saturation with water, and all conduits in wet outdoor locations or underground or in concrete or masonry in earth contact.

Luminaire (formerly lighting fixture): A complete lighting unit including parts to connect it to the power supply, and possibly parts to protect or distribute the light source. A lampholder, such as a porcelain socket, is not itself a luminaire.

Neutral conductor: The conductor connected to the neutral point of a system that is intended to carry current under normal conditions F17.

Open conductors: Individual conductors not contained within a raceway or cable sheathing, such as a typical service drop.

Outlet: The point on a wiring system at which current is taken to supply equipment. A receptacle or a box for a lighting fixture is an outlet; a switch is not an outlet.

Overcurrent: Any current in excess of the rating of equipment or conductor insulation. Overcurrents are produced by overloads, ground faults, or short circuits.

Overfusing: A fuse or breaker that has an overload rating greater than allowed for the conductor it is protecting.

Overload: Equipment drawing current in excess of the equipment or conductor rating and in such a manner that damage would occur if it continued for a sufficient length of time. Short circuits and ground faults are not overloads.

Panelboards: The "guts" of an electrical panel; the assembly of bus bars, terminal bars, etc., designed to be placed in a "cabinet." What is commonly called an electrical panel or load center is, by NEC terms, a panelboard mounted in a cabinet F16.

Power: There are 2 designations for AC electrical power. Apparent power (input) is expressed in V x A. True power (useful output) is expressed in watts.

Service: The conductors and equipment providing a connection to the utility F2.

Service drop: The overhead conductors supplied by the utility F2.

Service entrance conductors: The conductors from the service point to the service disconnect.

Service equipment: The equipment at which the power conductors entering the building can be switched off to disconnect the premises' wiring from the utility power source. A meter can be a part of or separate from the service equipment.

Service lateral: Underground service entrance conductors.

Service point: The connection or splice point at which the service drop and service entrance meet–it is the handoff between the utility and the customer.

Short circuit: A direct connection of current-carrying conductors without the interposition of a load, resulting in high levels of current.

Short Circuit Current Rating (SCCR): The amount of current that panelboards and switchboards must be able to carry during a short circuit condition without sustaining damage. See *interrupting rating*.

Snap switch: A typical wall switch, including 3-way and 4-way switches.

Ufer: a concrete-encased grounding electrode, named after the developer of the system, Herbert Ufer F6.

Unit switch: A switch that is an integral part of an appliance.

Within sight (also written as "in sight"): Visible, unobstructed, and not more than 50 ft. away.

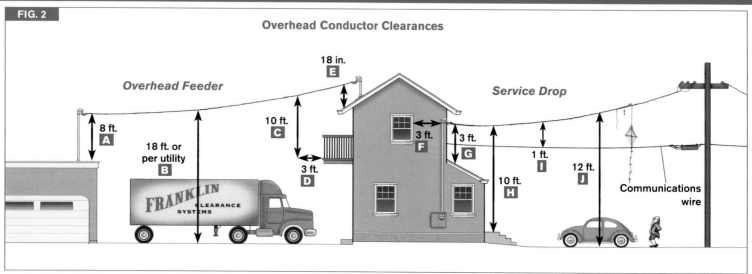

FIG. 2

Overhead Conductor Clearances

Overhead Feeder

Service Drop

18 in. **E**

10 ft. **C**

3 ft. **D**

8 ft. **A**

18 ft. or per utility **B**

FRANKLIN CLEARANCE SYSTEMS

3 ft. **F**

3 ft. **G**

1 ft. **I**

12 ft. **J**

10 ft. **H**

Communications wire

OVERHEAD SERVICE DROP CLEARANCES

Service drop conductors typically have no outer jacket for physical protection and no overload protection at their source. They are protected by isolation and proper clearances. The codes specify minimum clearances, and the serving utility may have different rules that override the code. Check with your local jurisdiction to determine any variations from the standard clearances below.

Vertical above Roof F2   09 IRC   11 NEC

☐ <4-in-12 slope: min 8ft **A** EXC_____ [3604.2.1]   {230.24A}
   • 3ft OK if roof area guarded or isolated _____[n/a] {230.24AX5}[1]
☐ ≥4-in-12 slope: min 3ft **G** EXC_____ [3604.2.1X2] {230.24AX2}
   • 18in OK for ≤4ft over eave **E** _____ [3604.2.1X3] {230.24AX3}
☐ Maintain req'd distance above roof for 3ft past
   edge EXC _____ [3604.2.1]   {230.24A}
   • Edge clearance above roof is not req'd when
   attached to side of building_____[3604.2.1X4] {230.24AX4}

Vertical above Grade F2   09 IRC   11 NEC

☐ 10ft above final grade to lowest point of drip loop _ [3604.2.2] {230.24B1}
☐ Area accessible only to pedestrians: 10ft **H** _____ [3604.2.2] {230.24B1}
☐ General above grade & driveways: 12ft **J** _____ [3604.2.2] {230.24B2}
☐ Above roads or parking areas subject
   to truck traffic: 18ft **B** _____ [3604.2.2] {230.24B4}
☐ Any direction from swimming pool water: 22½ft ____ [4103.5]   {680.8A}

Openings & Communication Wires F2   09 IRC   11 NEC

☐ Vertical above decks & balconies: 10ft **C** _____[n/a] {230.9B}
☐ From side of area above decks & balconies: 3ft **D** __ [3604.1] {230.9A}
☐ Below or to sides of openable window: 3ft **F** _____ [3604.1] {230.9A}
☐ Communications wire ≥12in in parallel power wires **I** ___[n/a] {800.44A4}

The clearances from windows and doors apply to open conductors and not to conductors contained inside a raceway or a cable with an overall outer jacket. The codes do not have a requirement for minimum clearance of open conductors above a window. Check to see if your local utility has a requirement.

SERVICE ENTRANCE CONDUCTORS

The connection between the service drop or lateral and the permanently installed building wiring is typically considered the "service point"–the handoff from the utility to the customer. From that point to the service equipment, the conductors are referred to as service entrance conductors. Though the utility does not have exclusive control of these conductors, they may still have jurisdiction over them, including the size of conduits and the placement of metering equipment.

General   09 IRC   11 NEC

☐ Wire size for SFD per **T10** _____ [T3603.1] {T310.15B7}
☐ Min wire size for SFD 4AWG Cu or 2AWG Al **T10** _ [T3603.1] {T310.15B7}
☐ Conductors & cables exposed to sunlight L&L as sunlight-resistant
   or covered w/ material L&L as sunlight-resistant_____ [3605.6] {310.10D}
☐ Identify (white marking or tape) neutral at both ends _ [3307.1] {200.6B}
☐ Service heads/goosenecks above attachment
   point EXC _____ [3605.9.3] {230.54C}
   • Attachment within 24in OK when necessary ___ [3605.9.3X] {230.54CX}
☐ No branch circuits or feeders in same
   raceway w/ service conductors _____ [3601.4] {230.7}
☐ Form drip loop in conductors _____ [3605.9.5] {230.54F}
☐ Individual open conductor insulating supports min 2in
   from building surfaces _____ [n/a] {230.51C}

Service Entrance Cables   09 IRC   11 NEC

☐ Protect SE cables subject to damage w/ metal
   conduit, PVC-80, EMT, or other approved
   means **F58,59,63**_____ [3605.5] {230.50B}
☐ Secure SE cable every 30in & 12in from terminations [3605.7] {230.51A}
☐ Raintight service head or taped gooseneck req'd _ [3605.9.2] {230.54B}
☐ Seal SE cable to prevent water entry to box _____ [3605.9.6] {230.54G}

Service Riser   09 IRC   11 NEC

☐ Wiring method listed for electrical (no plumbing pipe) [3605.2] {230.43}
☐ Suitable for wet location if exposed to weather _____ [3605.8] {230.53}
☐ Overhead raceway req's raintight service head ___ [3605.9.1] {230.54A}
☐ Brace riser to utility or local specifications _____ [3604.5] {230.28}
☐ Only power conductors on service risers–no CATV__ [3604.5] {230.28}
☐ Size raceway to max 40% fill **T17–T22** _____ [3904.6] {9-T1&T4}
☐ Size raceway per utility _____ [utility] {utility}

Aside from code issues, utility company rules and standards must be followed. Most utilities publish their gas and electrical service requirements or post them online. The following items are not in the codes, and you should consult with your local utility to comply with their rules on these issues.

Meter Base(s)

☐ Too close to gas meter
☐ Height incorrect
☐ Barrier post (bollard) needed to protect meter from vehicles on driveway
☐ Not readily accessible to meter readers

Service Entrance Conductors

☐ Insufficient conductor length at service head
☐ Insufficient clearance to communication lines
☐ Insufficient clearance above windows
☐ Height above standing surface (roof deck) too low
☐ Trees under service drop
☐ Customer performing own cutover from old service to new

SERVICE PANELS

The term "service equipment" refers to the switches, circuit breakers, or fuses that disconnect power from the utility at the customer's end of the service conductors. A meter is not considered service equipment, though it is sometimes in the same enclosure as the service equipment. As with all electrical equipment that might require access for maintenance, examination, or repair, sufficient working space must be maintained around service equipment.

General   09 IRC   11 NEC

☐ Enclosure L&L as suitable for service equipment __ [3601.6.1] {230.66}
☐ Max 6 disconnects to shut off power_____ [3601.7] {230.71}
☐ Service disconnects labeled as such_____ [3601.6.1] {230.70B}
☐ In multiple-occupancy building, each occupant
   must have ready access to disconnect EXC_____ [3601.6.2] {230.72C}
   • OK for management to have only access to
   service disconnect supplying >1 occupancy _____ [n/a] {230.72CX}
☐ Max height of breaker 6ft 7in _____ [4001.6] {240.24A}
☐ Provide working space **F3**_____ [3405.2] {110.26}

WORKING SPACE

Working space around equipment is essential for worker safety. These requirements apply to any electrical equipment that might require examination, adjustment, servicing, or maintenance while energized. The spaces around electrical equipment should not be used for storage.

Working Space F3	09 IRC	11 NEC
☐ Front working clearance min 36in deep	[3405.2]	{110.26A1}
☐ Distance measured from face of enclosure or live parts	[3405.2]	{110.26A1}
☐ Work space extends to floor EXC	[3405.2]	{110.26A3}
• Related equipment may extend 6in beyond panel front	[3405.2]	{110.26A3}
☐ Clear width min 30in wide or width of equipment	[3405.2]	{110.26A2}
☐ Panel need not be centered in space, hinged doors must be openable at least 90°	[3405.2]	{110.26A2}
☐ Working space not to be used for storage	[3405.4]	{110.26B}
☐ Illumination req'd for all indoor panels	[3405.6]	{110.26D}
☐ Min headroom for service & panels 6½ft	[3405.7]	{110.26A3}

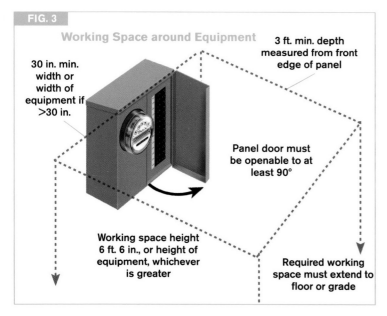

FIG. 3

Working Space around Equipment

30 in. min. width or width of equipment if >30 in.

3 ft. min. depth measured from front edge of panel

Panel door must be openable to at least 90°

Working space height 6 ft. 6 in., or height of equipment, whichever is greater

Required working space must extend to floor or grade

SEPARATE BUILDINGS

Care must be taken to avoid objectionable currents on the grounding paths between buildings supplied by a common service. Install separate insulated neutral conductors, rather than using the grounding conductors as neutrals. The IRC does not address outside feeders and separate buildings except for the rules on grounding.

Outside Feeders	11 NEC
☐ Trees may not support overhead conductors	{225.26}
☐ Overhead feeder height rules same as services F2	{225.18&19}
☐ Provide proper cover for buried cable or conduit F5,T1	{300.5A}
☐ Each building or structure req's GES EXC F4	{250.32A}
• Building or structure w/ only 1 branch circuit w/ EGC	{250.32AX}
☐ Multiwire circuit considered 1 circuit for above rule	{250.32AX}
☐ Seal underground raceways where entering building	{225.27}²
☐ Max one feeder or branch circuit to each building	{225.30}
☐ Max one feeder or branch circuit back to original building	{225.30}³
☐ Disconnect req'd at each building F4	{225.31}
☐ Disconnect must be rated as service equipment EXC	{225.36}
• Garages or outbuildings snap switches or 3-ways OK	{225.36X}
☐ Do not bond neutral to EGC or enclosure in subpanel (4-conductor feed req'd) EXC	{250.32B}
• Existing installations to separate buildings w/ no continuous metal paths, e.g. metal water pipe, etc. between the 2 structures	{250.32BX}

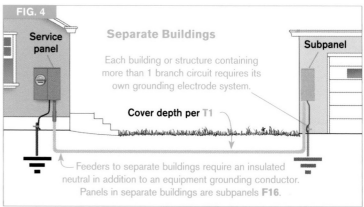

FIG. 4

Separate Buildings

Service panel

Subpanel

Each building or structure containing more than 1 branch circuit requires its own grounding electrode system.

Cover depth per T1

Feeders to separate buildings require an insulated neutral in addition to an equipment grounding conductor. Panels in separate buildings are subpanels F16.

MULTI-METER SERVICES

Services to two-family and multi-family dwellings might come to a multi-meter panel, or to a "hot gutter" with splices ahead of any overcurrent protection. See **p. 11** for bonding requirements on such services

Multi-Meter Services

		09 IRC	11 NEC
☐ Only 1 service per building		[3601.2]	{230.2}
☐ Provide each occupant access to service disconnect		[3601.6.2]	{230.72C}
☐ Bonding req'd at hot gutters F11,12		[3609.2]	{250.92A}
☐ Service disconnects grouped in 1 location		[3601.7]	{230.71A}
☐ Service conductors may not pass through interior of 1 building to another building		[3601.3]	{230.3}

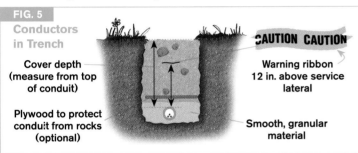

FIG. 5
Conductors in Trench

Cover depth (measure from top of conduit)

CAUTION CAUTION

Warning ribbon 12 in. above service lateral

Plywood to protect conduit from rocks (optional)

Smooth, granular material

TEMPORARY WIRING

Safety is the highest priority during construction, and GFCI protection is required for all 120V receptacles on construction sites. Some jurisdictions allow a limited number of temporary circuits from a service installation prior to the rough wiring stage (before weather protection).

Temporary Wiring

	11 NEC
☐ Allowed only during construction, repair, remodeling, & similar	{590.3A}
☐ Service conductor clearances same as permanent F2	{590.4A}
☐ Support & brace pole to utility specifications	{utility}
☐ Provide overcurrent protection for branch circuits	{590.4C}
☐ No receptacles on branch circuits supplying temporary lighting	{590.4D}
☐ All multiwire circuits req handle ties	{590.4E}
☐ Lampholders req guards	{590.4F}
☐ Splices of NM to NM or MC to MC OK w/o splice box	{590.4G}
☐ Protect cords & cables from accidental damage	{590.4H}
☐ GFCI req'd all 125 15, 20, & 30A temporary receptacles EXC	{590.6A1}
• Listed GFCI cord-sets OK on existing permanent receptacles	{590.6A2}[4]

UNDERGROUND WIRING

Underground wiring methods include individual insulated conductors, cables rated for underground installation, and raceways. The most common method is PVC conduit. If there is a significant difference in elevation between the ends of an underground raceway, it may be necessary to install a pull-box for drainage near the downhill end.

General

		09 IRC	11 NEC
☐ Burial depth must provide cover per T1,F5		[3803.1]	{300.5A}
☐ Warning ribbon in trench 12in above service laterals F5		[3803.2]	{300.5D3}
☐ Direct-buried cables or conductors must be protected by enclosures or raceways from req'd burial depth or 18in (whichever is less) to termination above grade or 8ft high (whichever is less) F52		[3803.3]	{300.5D1}
☐ Protect conductors & cables emerging from grade w/ RMC, IMC, PVC-80, or equivalent F52		[3803.3]	{300.5D4}
☐ OK to splice or tap direct-buried conductors w/o boxes w/ splicing means listed for the purpose		[3803.4]	{300.5E}
☐ Backfill smooth granular material–no rocks F5		[3803.5]	{300.5F}
☐ Boards or sleeves for protection where necessary F5		[3803.5]	{300.5F}
☐ Seal underground raceway entries (vapor protection)		[3803.6]	{300.5G}
☐ Bushing req'd between underground cables or individual conductors & protective conduit F52		[3803.7]	{300.5H}
☐ All conductors of circuit in same trench or raceway		[3803.8]	{300.5I}
☐ Allow provision for earth movement (settlement or frost) using "S" loops, flexible connections, &/or expansion fittings)		[3803.9]	{300.5J}

TABLE 1	MIN. COVER REQUIREMENTS IN TRENCH [T3803.1] & {300.5}				
Cover	UF Cable	Rigid Metal	PVC	GFCI ≤20A Circuit	≤30V
General	24 in.	6 in.	18 in.	12 in.	6 in.
2 in. concrete	18 in.	6 in.	12 in.	6 in.	6 in.
Under building	0[A]	0	0	n/a	n/a
4 in. slab, no vehicles	18 in.	4 in.	4 in.	6 in.	6 in.
Street	24 in.	24 in.	24 in.	24 in.	24 in.
Driveway	18 in.	18 in.	18 in.	12 in.	18 in.

A. MC cable identified for direct burial also OK in 2011 NEC

SERVICE & FEEDER LOAD CALCULATIONS

The calculation methods in the codes take into account that not all of the possible electrical loads will be used at the same time. Each of the calculation methods allows the use of "demand factors." The "long form," shown below, is the most common calculation method. For information on multifamily load calculations, refer to the Code Check website at **www.codecheck.com**.

Load Calculation Steps (long form) T2   **11 NEC**

1. Determine the sq-ft area of the residence & multiply by 3W (exclude garage & covered patios) _____ {220.12}
2. Min of 2 small-appliance circuits at 1,500W each _____ {220.52A}
3. Each additional small appliance circuit at 1,500W _____ {220.52A}
4. Minimum 1 laundry circuit at 1,500W _____ {220.52B}
5. Enter total of appliance circuits & general lighting _____ {220.42}
6. First 3,000W counted at 100% (carries to right column) _____ {T220.42}
7. Subtract 3,000 from amount in line 5 & enter difference in middle column. Multiply the middle column amount by 35% & enter in right column _____ {T220.42}
8. Range loads are calculated at nameplate rating. If a single range is >8,000W & <12,000W, it still counts as 8,000W (8kW); >12,000W, add 5% of each additional 1,000W of nameplate load. The nameplates of a counter-mounted range & up to 2 wall ovens can be added together & computed as if they were 1 range. Enter in right column _____ {220.55}
9. Enter dryer circuit at 5,000W (or nameplate rating if greater) _____ {220.54}
10. Enter larger of fixed space heating or AC load _____ {220.60}
11–18. Enter the nameplate ratings of appliances that are fixed in place. For appliances rated in amps, multiply amps times voltage to determine watts. If nameplate ratings unknown, use estimates in **T4** _____ {220.53}
19. Enter total load of fixed appliances _____ {220.53}
20. If there are <4 fixed appliances, enter the number from line 19 in the right column _____ {220.53}
21. If there are ≥4 fixed appliances, multiply line 19 by 75% & enter in the right column _____ {220.53}
22. Add 25% of the largest motor load. Skip this step if a nameplate rated AC is the largest load since the number has already been factored into the nameplate min conductor ampacity _____ {220.18A}
23. Add the numbers in the third column _____ {220.40}
24. Divide line 23 by 240 to find req'd min amperage _____ {220.40}

TABLE 2	LOAD CALCULATIONS [T3704.2(1)] & {220.40}		
General Lighting & Receptacle Loads:			
1	Sq. ft. × 3W		
Small Appliance & Laundry Loads			
2	2 small appliance circuits	3,000	
3	Additional small appliance		
4	Laundry circuit	1,500	
5	Subtotal general light, small appliance & laundry		
6	First 3,000W @ 100%	3,000	3,000
7	Balance @ 35%	× .35	=
Special Appliance Loads:			
8	Range	8,000 up to 12kW nameplate	
9	Dryer	5,000 (or nameplate if >)	
10	Heating or AC @ 100%		
Appliances Fastened in Place:			
11	Water heater		
12	Microwave		
13	Dishwasher		
14	Compactor		
15	Disposer		
16	Attic fan		
17	Spa–per manu.		
18	Other		
19	Subtotal		
20	If <4 appliances, enter subtotal @100% OR		
21	If ≥4 appliances, enter subtotal × 75%		
22	Largest motor × 25%		
23	Total load		
24	Total load ÷ 240V = SERVICE AMPS		

Size Requirements–General

	09 IRC	11 NEC
☐ Min size for SFD 100A _____	[3602.1]	{230.79C}
☐ Service conductors adequate for load served_____	[3602.1]	{230.42}
☐ Feeders adequate for load served _____	[3701.2]	{215.2A1}
☐ Branch circuits adequate for load served _____	[3701.2]	{210.19A1}

*The "optional" method is simpler and can also be used to determine if an existing service is adequate for expansion. In the NEC, these methods apply to both services and feeders. In the IRC, the "long form" method **T2** is used for feeders per section E3704, and the "optional" method **T3** is used for services per section E3602. NEC 220.83 provides a specific method for evaluating the adequacy of an existing service for new air conditioning loads.*

TABLE 3	MIN. SIZE OF ELECTRICAL SERVICE [T3602.2] & {220.82}		
1. Indoor sq. ft. × 3VA/ft.			
2. Min. 2 small appliance circuits @ 1,500VA each	3,000		
3. Laundry circuit @ 1,500VA	1,500		
4. Nameplate VA of fixed appliances:			
Dryer @ 5,000VA			
Oven(s)			
Cooktop			
Water heater			
Dishwasher			
Disposer			
Other			
5. Subtotal			
6. First 10,000VA @ 100%	10,000	10,000	
7. Balance @ 40% (subtract line 6 from line 5)	× .40	=	
8. Largest of heating or cooling load:			
8a. Nameplate rating(s) of air-conditioning & cooling equipment OR			
8b. Heat pump nameplate if no supplemental electric heat OR			
8c. Continuous electric thermal storage @ nameplate rating OR			
8d. 100% of heat pump nameplate rating plus 65% of supplemental electric heat or central electric heat OR			
8e. Space heaters @ 65% of nameplate rating if < 4 units OR			
8f. Space heaters @ 40% of nameplate rating if ≥ 4 units			
9. Total load in VA			
10. Divide by 240 = minimum service rating			

TABLE 4	TYPICAL APPLIANCE LOADS

Use actual nameplate ratings when known. This table is for estimating purposes when appliances are not yet specified.

Appliance	Typical load (watts)
Central AC or heat pump	1,800 per ton
Dishwasher	1,200
Food waste disposer	900
Trash compactor	1,200
Microwave	1,500
Central furnace	1,000
Central vacuum	1,500
Electric clothes dryer	5,000
Water heater	4,500
Electric cooktop	3,600
Single wall oven	4,800
Double wall oven	8,000
Pool pump	2,000
Well pump	2,000

Optional Method (Short Form) 11 NEC

1. 3W per ft (exclude garage & covered patios) _____ {220.82B1}
2. Min 2 small-appliance circuits at 1,500W each, each additional small appliance circuit at 1,500W _____ {220.82B2}
3. Min one laundry circuit at 1,500W _____ {220.82B2}
4. Nameplate ratings of fixed appliances (see **T4** if ratings not known); these include the full nameplate rating of ranges & ovens w/o applying the reductions allowed in the "long form" method_____ {220.82B3}
5. Enter sum of items 1–4 _____ {220.82B}
6. 100% of first 10,000VA _____ {220.82B}
7. Subtract line 6 from line 5, multiply by 40% & enter in right column _____ {220.82B}
8. Determine the largest of the heating or cooling load. When using the nameplate rating of heat pumps or air conditioning, multiply the "minimum circuit ampacity" times the voltage (240). If only the size (tonnage) is known, refer to **T4**_____ {220.82C}
9. Add the numbers in the right column & enter total _____ {220.82A}
10. Divide by 240 = amperage

GROUNDING ELECTRODES

Grounding electrodes are metal conducting objects through which a direct connection to earth is established. These electrodes provide a path for lightning and help reduce electrical noise on communications equipment. The most common grounding electrodes in residential construction are metal underground water piping, ground rods, and concrete-encased electrodes.

Grounding Electrode System (GES) F6   09 IRC   11 NEC
- ☐ Use all electrodes in F6 when present on premises__ [3608.1]   {250.50}
- ☐ Electrodes bonded together form a single system F6 __ [3608.1]   {250.50}
- ☐ Size electrode bonding conductors per GEC rules __ [3610.1]   {250.53C}
- ☐ Underground gas pipe not OK as electrode _____ [3608.6]   {250.52B1}

Water Pipe   09 IRC   11 NEC
- ☐ Metal water pipe if ≥10ft in direct contact w/ soil__ [3608.1.1]   {250.52A1}
- ☐ Bond around water meters, filters, etc. _____ [3608.1.1.1]   {250.53D1}
- ☐ Water pipe cannot be sole electrode _____ [3608.1.1.1]   {250.53D2}
- ☐ Metal well casing that is not bonded to metal pipe (e.g., plastic water service from well) OK as electrode [3608.1.1]   {250.52A8}

Pipes & Rods   09 IRC   11 NEC
- ☐ Rods min 8ft in contact w/ soil F6 _____ [3608.1.4.1]   {250.53G}
- ☐ Pipe electrodes min 3/4in diameter _____ [3608.1.4]   {250.52A5}
- ☐ Unlisted ground rods min 5/8in diameter _____ [3608.1.4]   {250.52A5}
- ☐ Listed rods min 1/2in diameter _____ [3608.1.4]   {250.52A5}
- ☐ Drive rods vertical & fully below grade EXC ____ [3608.1.4.1]   {250.53G}
  - If bedrock encountered, rod may be buried horizontally 2 1/2ft deep or driven at 45° angle _____ [3608.1.4.1]   {250.53G}
  - Clamp above grade OK if protected F6-10 ___ [3608.1.4.1]   {250.53G}
- ☐ If rod resistance >25 ohms, install 2nd rod min 6ft from first & bond to 1st rod _____ [3608.4] {250.53A2X}

Recommended spacing 2X rod length, i.e., 16ft

Concrete-Encased Electrode F6   09 IRC   11 NEC
- ☐ Ufer = 20ft #4 or larger rebar near bottom of footing or 20ft 4AWG or larger Cu wire near bottom of footing [3608.1.2]   {250.52A3}
- ☐ Ufer must be used if present during construction____ [3608.1]   {250.50}
- ☐ Ufer not req'd in existing building if concrete would have to be disturbed to gain access _____ [3608.1X]   {250.50X}
- ☐ Ufer concrete encasement min 2in_____ [3608.1.2]   {250.52A3}
- ☐ OK to bond sections of rebar w/ ordinary steel tie wires _____ [3608.1.2]   {250.52A3}
- ☐ Where multiple concrete-encased electrodes are present, only 1 req'd to be bonded to GES _____ [3608.1.2]⁵   {250.52A3}
- ☐ Metal building frame OK as electrode if bonded to Ufer or if ≥ 10ft of steel in contact w/ earth w/ or w/o concrete encasement _____[n/a]   {250.52A2}

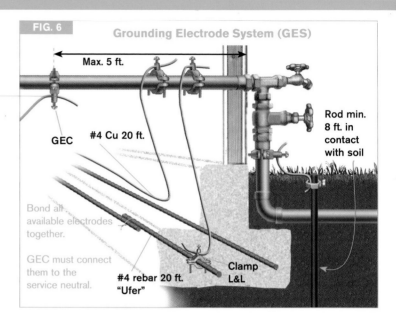

FIG. 6   Grounding Electrode System (GES)

Max. 5 ft.

GEC   #4 Cu 20 ft.

Rod min. 8 ft. in contact with soil

Bond all available electrodes together.

GEC must connect them to the service neutral.

#4 rebar 20 ft. "Ufer"

Clamp L&L

GROUNDING ELECTRODE CONDUCTORS (GECs)

GECs connect the system of metal grounding electrodes in earth to the electrical system. They must have adequate size and protection to withstand environmental and electrical forces imposed on them. Individual conductors can be run to each electrode of the grounding electrode system or a single conductor can be run to one of them or to the conductor that bonds the electrodes to each other.

Locations   09 IRC   11 NEC
- ☐ GEC must connect to EGCs, service entrance enclosures, service neutral, & grounding electrodes_____ [3607.4]   {250.24D}
- ☐ Connect to service neutral anywhere from service point to bonded neutral in service disconnect _____ [3607.2]   {250.24A1}
- ☐ Bare Al not OK in masonry or earth _____ [3610.2]   {250.64A}
- ☐ Where outside, no Al ≤18in of earth _____ [3610.2]   {250.64A}
- ☐ Connection to metal water pipe that is part of GES not >5ft after water entry to building F6_____ [3608.1.1]   {250.68C1}

Protection F7-10   09 IRC   11 NEC
- ☐ 8AWG must be protected by raceway or armor F8,9 [T3603.1]   {250.64B}
- ☐ 6AWG OK unprotected if not subject to damage & following building contour F7 _____ [T3603.1]   {250.64B}
- ☐ Bond each end of metal raceway enclosing GEC F9 [T3603.1]   {250.64E}

GROUNDING ELECTRODE CONDUCTORS (CONT.)

Size	09 IRC	11 NEC
☐ Size per service conductor size T5 EXC _____ [3603.4]		{250.66}
• 6AWG Cu largest size GEC needed if ending at rod [T3603.1]		{250.66A}
• 4AWG Cu largest size GEC needed if ending at Ufer [T3603.1]		{250.66B}

Connections	09 IRC	11 NEC
☐ No splices between service & GES EXC _____ [3610.1]		{250.64C}
• Listed irreversible compression connectors or exothermic welding OK _____ [3610.1X]		{250.64C}
☐ GEC can connect to any electrode of GES _____ [3610.1]		{250.64F}
☐ Buried clamps L&L for direct burial (marked "DB") F6 [3611.1]		{250.70}
☐ Cu water tubing clamps L&L for Cu tubing _____ [3611.1]		{250.70}
☐ Ufer clamps L&L for rebar & encasement F6 _____ [3611.1]		{250.70}
☐ Strap-type clamps suitable only for indoor telecommunications _____ [3611.1]		{250.70}
☐ Max 1 conductor per clamp unless listed for more___ [3611.1]		{250.70}
☐ Connections must be accessible EXC F6 _____ [3611.2]		{250.68A}
Buried or encased connections F6 _____ [3611.2]		{250.68AX}

Note: Rebar can be brought through the top of a foundation in a protected location, such as the garage, to provide an accessible point for the GEC to attach to the Ufer. The GEC can also be brought into the foundation and connect to the Ufer with L&L clamps or by exothermic welding.

TABLE 5	GEC SIZING [T3603.1] & {T250.66}	
Cu Service Wire AWG	**Al Service Wire AWG**	**GEC Cu AWG**
≤2	≤1/0	8
1 or 1/0	2/0 or 3/0	6
2/0 or 3/0	4/0 or 250kcmil	4
4/0–350kcmil	>250–500kcmil	2
>350–600kcmil	>500–900kcmil	1/0

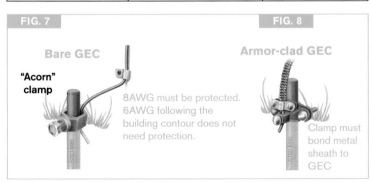

FIG. 7

Bare GEC

"Acorn" clamp

8AWG must be protected. 6AWG following the building contour does not need protection.

FIG. 8

Armor-clad GEC

Clamp must bond metal sheath to GEC

FIG. 9

GEC in Metal Raceway

FIG. 10

GEC in PVC

Conductive protection must be bonded at both ends, making PVC a simpler solution.

EQUIPMENT GROUNDING CONDUCTORS (EGCs)

EGCs limit the voltage on equipment enclosures and provide a path for fault current. Without EGCs, the conductive frame of an appliance could remain energized if there is a fault from an ungrounded "hot" conductor. Equipment grounding provides a low-impedance path so the overcurrent device will open the circuit. The equipment grounding system has a completely different purpose from the earth grounding system. In fact, earth plays no part in helping to clear faults.

Equipment Grounding Conductors	09 IRC	11 NEC
☐ EGC must provide effective ground-fault current path [3908.4]		{250.4A5}
☐ Earth is not an effective ground-fault current path ___ [3908.5]		{250.4A5}
☐ Size EGCs per T6 _____ [3908.12]		{250.122A}
☐ RMC, IMC, EMT, AC cable armor, electrically continuous raceways, & surface metal raceways OK as EGC ___ [3908.8]		{250.118}
☐ Wire EGCs can be bare, covered, or insulated F16 _ [3908.8]		{250.118}
☐ Insulation on EGC green or green w/ yellow stripes _____ [n/a]		{250.119}
☐ EGC >6AWG OK to strip bare for entire exposed length or use green tape or labels at the termination of the wire _____ [n/a]		{250.119A}
☐ FMC & LFMC OK as EGC for non-motor circuits in combined lengths to 6ft w/ grounding fittings F60,61 _____ [3908.8.1&2]		{250.118}
☐ Remove paint from threads & other contact surfaces for field-installed equipment such as ground terminal bars _____ [3908.17]		{250.12}
☐ EGCs must be run w/ other conductors of circuit EXC [3406.7]		{300.3B}
• Replacing nongrounding receptacles (see **p.29**) _____ [n/a]		{250.130C}
☐ Neutral not to be used for grounding equipment EXC [3908.7]		{250.142B}
• Existing ranges & dryers _____ [n/a]		{250.142BX1}

TABLE 6	EQUIPMENT GROUNDING CONDUCTORS (EGCs) [T3908.12] & {T250.122}	
Size in Amps of Breaker or Fuse Protecting Circuit	**AWG Size of Cu EGC**	**AWG Size of Al EGC**
15	14	12
20	12	10
30–60	10	8
70–100	8	6
110–200	6	4
400	3	1

BONDING

Bonding ensures electrical continuity to limit voltage potential between conductive components. On the **line side** (ahead of the main disconnect **F15**), it provides a path back to the utility transformer for faults on service conductors and to limit voltage potential to other systems, such as telephones or cable TV. On the **load side** (after the main overcurrent protection **F15**), bonding and equipment grounding provide a path to clear faults and protect against shocks.

Bonding & Equipment Grounding Methods   09 IRC   11 NEC

- ☐ Use listed connectors, terminal bars, exothermic welding, machine screws engaging 2 threads or secured w/ nut, or thread-forming machine screws engaging 2 threads—no sheet metal or drywall screws _____ [3908.15][6]   {250.8A}
- ☐ Connections may not depend solely on solder _____ [3908.13]   {250.8B}
- ☐ Clean nonconductive coatings from contact surfaces [3908.17]   {250.12}

Line-Side Bonding F11,12,15   09 IRC   11 NEC

- ☐ Bond all service equipment, raceways, & cable armor [3609.2]   {250.92A}
- ☐ Bond metal GEC enclosures at each end_____ [T3603.1]   {250.64E}
- ☐ Threaded fittings OK for bonding service conduit _ [3609.4.2]   {250.92B2}
- ☐ Meyers hub OK for bonding service conduit **F11**__ [3609.4.2]   {250.92B2}
- ☐ Standard locknuts alone not sufficient on line side of service **F11** _____ [3609.4.3]   {250.92B2}
- ☐ Bonding locknuts OK if no remaining concentrics **F11** [3609.4.4]   {250.92B4}
- ☐ Jumpers req'd around concentric knockouts, or reducing washers on line side of service **F12,15**_____ [3609.4.4]   {250.92B4}[7]
- ☐ Service neutral can bond line-side equipment _____ [3609.4.1]   {250.142A}
- ☐ Size line-side bonding jumpers per **T5**_____ [3609.5]   {250.102C}
- ☐ Service enclosure main bonding jumper must connect enclosure, service neutral, & equipment grounds **F15** _____ [3607.5]   {250.24B}

Load-Side Bonding   09 IRC   11 NEC

- ☐ Bond any metal piping system capable of becoming energized, including hot & cold water & gas **F13** _ [3609.6&7]   {250.104}
- ☐ Size water pipe bonding per **T5** _____ [3609.6]   {250.104A1}
- ☐ Size gas pipe bonding per **T6**_____ [3609.7]   {250.104B}
- ☐ Bond metal well casings to EGC of pump motor_____[n/a]   {250.112M}

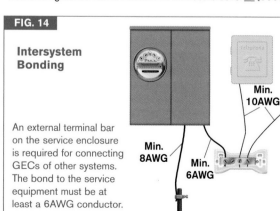

FIG. 13

Bonding Interior Piping

All interior piping systems capable of becoming energized must be bonded, & connecting them at a gas water heater provides an easy way to check for compliance.

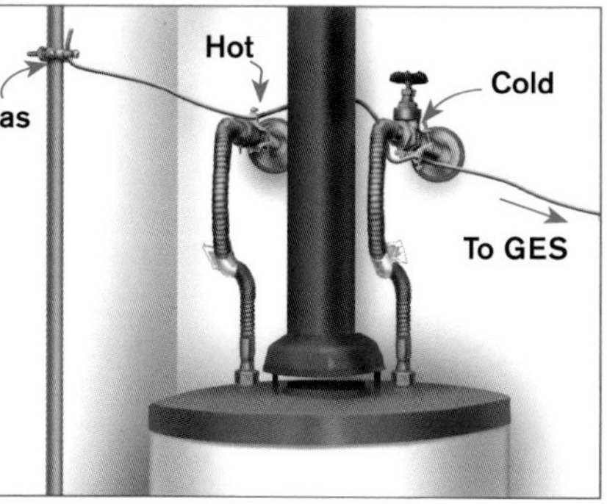

Hot   Cold   Gas   To GES

Intersystem Bonding F14   09 IRC   11 NEC

- ☐ Min 6AWG Cu bond to CATV or phone electrodes_____[3609.3][8]   {800.100D}
- ☐ Bond lightning protection system to GEC _____[n/a]   {250.106}
- ☐ Intersystem bonding access req'd external to service equipment & separate structure disconnecting means [3609.3]   {250.94}
- ☐ Must accept min 3 conductors & be terminal or bonding bar electrically connected to meter or service enclosure_____ [3609.3][8]   {250.94}
- ☐ Existing buildings raceway or GEC OK as bond point___ [n/a][8]   {250.94X}
- ☐ Bonding device not to interfere w/ enclosure cover__ [3609.3]   {250.94}

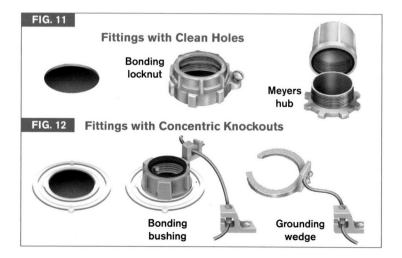

FIG. 11

Fittings with Clean Holes

Bonding locknut

Meyers hub

FIG. 12   **Fittings with Concentric Knockouts**

Bonding bushing   Grounding wedge

FIG. 14

Intersystem Bonding

An external terminal bar on the service enclosure is required for connecting GECs of other systems. The bond to the service equipment must be at least a 6AWG conductor.

Min. 10AWG

Min. 8AWG

Min. 6AWG

PANELBOARD & CABINETS

What is commonly called an "electrical panel" is referred to as a panelboard (NEC 408) inside a cabinet (NEC 312). See **p.5** for working space requirements.

Clearances & Location

	09 IRC	11 NEC
☐ No panels or OCPDs in clothes closet or bathroom _ [3405.4]		{240.24D&E}
☐ No panels or OCPDs over steps of a stairway _____ [3405.4][9]		{240.24F}
☐ OCPDs readily accessible & max height 6ft 7in _____ [3705.7]		{240.24A}

Enclosures

	09 IRC	11 NEC
☐ Enclosures weatherproof in wet or damp locations __ [3907.2]		{312.2}
☐ Surface-mounted wet or damp location metal enclosures min ¼in air gap between enclosure & wall _____ [3907.2]		{312.2}
☐ Equipment rated for dry or damp locations must be protected against damage from weather during construction___ [3404.4]		{110.11}
☐ Open knockouts & twistouts durably filled EXC _____ [3404.5]		{110.12A}
• Manu holes for mounting OK _____ [3404.6&3907.5]		{110.12A}
☐ Protect bus bars & other internal parts from contamination (paint or plaster) during construction_____ [3404.7]		{110.12B}
☐ Max setback in noncombustible wall ¼in _____ [3907.3]		{312.3}
☐ Flush (no setback) in combustible (wood-frame) wall [3907.3]		{312.3}
☐ Max plaster gap at side of flush mount panel ⅛in ___ [3907.4]		{312.4}
☐ Field labeling to distinguish each circuit from all others _ [3706.2]		{408.4}
☐ Labeling not based on transient conditions_____ [3706.2][10]		{408.4}
☐ Unused (spare) breakers labeled _____ [3706.2][11]		{408.4}

Grounding & Bonding

	09 IRC	11 NEC
☐ Bond neutral bar to enclosure & EGCs in service **F15** [3607.5]		{250.24B}
☐ Isolate neutrals in subpanels **F16**_____ [3607.2&3908.6]		{250.24A5}
☐ Grounding terminal bar req'd if wire EGCs present **F16** __[n/a]		{408.40}
☐ Continuity of neutral not to depend on enclosures [3406.11][12]		{200.2B}
☐ Each neutral conductor req's individual terminal_____ [3706.4]		{408.41}

OCPDs & Wiring

	09 IRC	11 NEC
☐ Panels req OCPD line side of bus **F15** _____ [3706.3]		{408.36}
☐ Breakers listed or classified AMI for panel _____ [3403.3]		{110.3B}
☐ Single-pole breakers w/ approved handle ties OK for 240V circuits **F16** _____[n/a]		{240.15B2}
☐ All multiwire circuits req handle tie or single handle [3701.5.1][13]		{210.4B}
☐ Handle tie req'd for 2 circuits to receptacles on same yoke [n/a]		{210.7B}
☐ All conductors of multiwire circuit must be grouped (wire ties or other means) inside panel EXC **F16** _ [3701.5.1][14]		{210.4D}
• Cable systems where grouping is obvious **F16** _ [3701.5.2][14]		{210.4DX}
☐ Backfed breakers secured in place _____ [3706.5]		{408.36D}
☐ Also applies to circuits from PV inverters _____[n/a]		{690.10E}
☐ Torque all breakers & terminals AMI _____ [3403.3]		{110.3B}
☐ Antioxidant on Al conductors AMI _____ [local]		{local}
☐ Secure each cable entering panel AMI **F15,16** _____ [3907.8]		{312.5C}
☐ Splices & taps in panels OK to 40% fill_____ [3907.1]		{312.8}
☐ Apply warning label to enclosure identifying power source of feed-through conductors_____[n/a]		{312.8}[15]

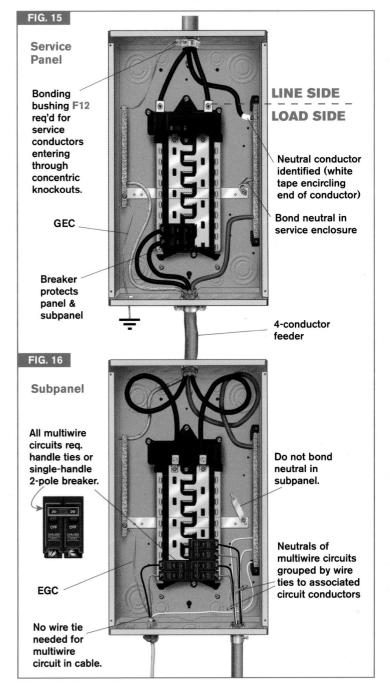

FIG. 15

Service Panel

Bonding bushing **F12** req'd for service conductors entering through concentric knockouts.

GEC

Breaker protects panel & subpanel

LINE SIDE

LOAD SIDE

Neutral conductor identified (white tape encircling end of conductor)

Bond neutral in service enclosure

4-conductor feeder

FIG. 16

Subpanel

All multiwire circuits req. handle ties or single-handle 2-pole breaker.

Do not bond neutral in subpanel.

EGC

Neutrals of multiwire circuits grouped by wire ties to associated circuit conductors

No wire tie needed for multiwire circuit in cable.

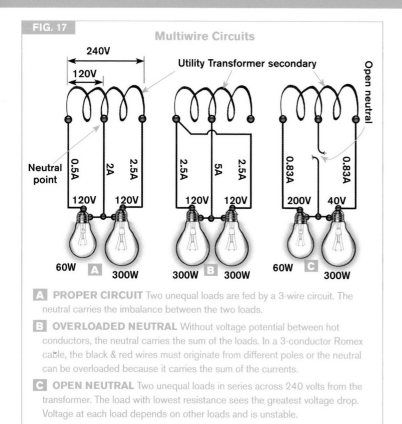

FIG. 17 — Multiwire Circuits

A PROPER CIRCUIT Two unequal loads are fed by a 3-wire circuit. The neutral carries the imbalance between the two loads.

B OVERLOADED NEUTRAL Without voltage potential between hot conductors, the neutral carries the sum of the loads. In a 3-conductor Romex cable, the black & red wires must originate from different poles or the neutral can be overloaded because it carries the sum of the currents.

C OPEN NEUTRAL Two unequal loads in series across 240 volts from the transformer. The load with lowest resistance sees the greatest voltage drop. Voltage at each load depends on other loads and is unstable.

3-WIRE EDISON CIRCUITS (MULTIWIRE) F17

Standard electrical services to one- and two-family dwellings originate at a utility transformer with two ungrounded "hot" conductors and a neutral derived from the center of the transformer's secondary coil, as depicted in F17. The neutral is connected to earth and is referred to as the "grounded" conductor. The neutral limits the voltage on either of the hot conductors to 120V to ground. If the neutral is broken or loose, voltages become erratic as in F17 C. TV sets, motors, and computers don't do well with fluctuating voltages. The utility company should be notified if there are signs of unstable voltage, such as incandescent bulbs growing brighter or dimmer as other loads change. Not only is the service to the house a "3-wire" circuit, but 120V branch circuits are often installed with shared neutrals, which are then known as multiwire circuits.

Multiwire Circuits (also see p.12)	09 IRC	11 NEC
☐ Hot conductors must originate from opposite poles ___	[3501]	{100}
☐ All conductors must originate from same panel _____	[3701.5]	{210.4A}
☐ Multiwire neutrals may not feed through devices such as receptacles (pigtail lead from neutral to device in box) _____	[3406.10.2]	{300.13B}
☐ Common neutral not OK for >1 multiwire circuit _____	[n/a]	{200.4}

AFCIs—ARC-FAULT CIRCUIT INTERRUPTERS

AFCIs provide fire protection by opening the circuit when an arcing fault is detected F18. They look similar to GFCI breakers F26, and AFCIs do provide some protection against shock hazards, though not at the level required for GFCIs. The 2008 NEC and 2009 IRC greatly expanded the areas that require AFCI protection. The time to plan for the AFCIs is during the rough wiring, so that separate cables are provided for the circuits requiring AFCI protection. Not all brands and models of AFCI are compatible with multiwire circuits.

Beginning January 1, 2008, all AFCIs were required to be "combination" type rather than the original "branch/feeder" type. Combination AFCIs provide a broader range of protection. Outlet types are mentioned in the codes, though at press time these were not yet available.

Acceptance of the AFCI code provisions varies widely by jurisdiction. Be sure to check with your local building department for their current AFCI requirements before beginning a wiring project.

AFCI Protection	09 IRC	11 NEC
☐ Combination-type AFCI req'd for 15A & 20A branch circuits supplying outlets in family rooms, dining rooms, living rooms, parlors, libraries, dens, bedrooms, sunrooms, recreation rooms, closets, hallways, & similar rooms or areas EXC _____	[3902.11][16]	{210.12A}
• Not req'd on individual circuit for central station alarm in RMC, IMC, EMT, or steel-armored cable (type AC) _____	[3902.11X2]	{210.12AX2}
☐ AFCI must protect entire branch circuit EXC _____	[3902.11]	{210.12A}
• OK to have protection at first outlet if wiring method between breaker & outlet is RMC, IMC, EMT, MC, or steel-armored cable (type AC) & metal outlet or junction boxes are used_____	[3902.11X1]	{210.12AX1}
☐ Replacement or extension of branch circuit wiring req's AFCI breaker at origin of replacement circuit or AFCI outlet device at first receptacle of existing branch circuit_____	[n/a]	{210.12B}[17]

FIG. 18 — Arc Fault

Loose connections at terminals are a common source of series arcs leading to electrical fires.

= CURRENT
= VOLTAGE

BOXES

Boxes must be large enough to contain all the conductors and devices inside them, and sufficient wire must be brought into the box to safely make up connections. Luminaires that are supported from boxes are generally designed so their connections will be made inside the box, rather than inside the fixture canopy. Device boxes are threaded for 6/32 screws used to mount switches and receptacles. Lighting outlet boxes provide 8/32 (for luminaires) or 10/24 screws (for listed paddle fan boxes).

General

	09 IRC	11 NEC
☐ Metal boxes must be grounded _____	[3905.2]	{314.4}
☐ Box & conduit body covers must remain accessible	[3905.10]	{314.29}
☐ Max ¼in setback from noncombustible surface F19_	[3906.5]	{314.20}
☐ Box extenders OK to correct excess setback _____	[3906.5]	{I314.20}
☐ Boxes flush w/ combustible surface F19_____	[3906.5]	{314.20}
☐ Plaster gap max⅛in for flush cover boxes F19 _____	[3906.6]	{314.21}
☐ Min 6in free conductor & 3in past box face _____	[3306.10.3]	{300.14}
☐ Luminaires only in boxes designed for luminaires EXC	[3905.6]	{312.27A}
• Wall sconces ≤6lb on device boxes w/ 2 #6 screws	[3905.6X]	{314.27A1X}
☐ Wall luminaire boxes marked w/ max weight if not 50lb	[3905.6][18]	{314.27A1}
☐ Ceiling luminaire boxes req 50lb rating F21 _____	[3905.7][18]	{314.27A2}
☐ Ceiling luminaires >50lb req independent support _	[3905.7][18]	{314.27A2}
☐ Smoke alarms OK on device boxes w/ 2 #6 screws _____[n/a]		{314.27DX}
☐ Paddle fans req L&L paddle fan box F42_____	[3905.9]	{314.27C}
☐ Boxes must be securely supported _____	[3906.8]	{314.23}
☐ PVC & EMT not OK for box support _____	[3906.8.5]	{314.23E}
☐ PVC & EMT OK for conduit body support_____	[3906.8.5]	{314.23E}
☐ Wet location boxes & conduit bodies listed for wet _	[3905.12]	{314.15}
☐ Damp or wet location boxes must keep out water __	[3905.12]	{314.15}

Box Fill

	09 IRC	11 NEC
☐ Size sufficient to provide free space for conductors	[3905.13]	{314.16}
☐ Standard metal boxes per code tables _____	[3905.13.1.1]	{314.16A1}
☐ Include volume of marked mud rings & extensions	[3905.13.1]	{314.16A}
☐ Plastic boxes have volume marking _____	[3905.13.1.2]	{314.16A2}
☐ 4in (6cu in) pancake OK only end of 14/2 run ___	[3905.13.2]	{314.16B}
☐ 18cu in box too small for 3 12/2 Romex T8,F20 __	[3905.13.2]	{314.16B}

TABLE 7	BOX FILL WORKSHEET [3905.13.2] & {314.16B}		
Item	Size	#	Total
#14 conductors exiting box	2.00		
#12 conductors exiting box	2.25		
#10 conductors exiting box	2.50		
#8 conductors exiting box	3.00		
#6 conductors exiting box	5.00		
Largest grounding conductor—count only one		1	
Devices—2× times connected conductor size			
Internal clamps—1 based on largest wire present		1	
Fixture fittings—1 for each type based on largest wire			
		TOTAL	

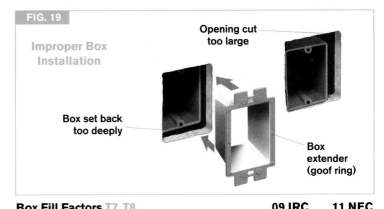

FIG. 19

Improper Box Installation

Opening cut too large

Box set back too deeply

Box extender (goof ring)

Box Fill Factors T7, T8

	09 IRC	11 NEC
☐ Count each conductor exiting box EXC_____	[3905.13.2.1]	{314.16B1}
• EGCs from luminaires or up to 4 conductors <14AWG from luminaires w/ domed canopies _____	[3905.13.2.1X]	{314.16B1X}
☐ Unbroken conductors passing through box count as only 1 conductor EXC _____	[3905.13.2.1]	{314.16B1}
• Looped unbroken conductors >12in count as 2	[3905.13.2.1]	{314.16B1}
☐ Do not count pigtailed conductors to devices _	[3905.13.2.1]	{314.16B1}
☐ All internal clamps count as 1, based on largest conductor in box_____	[3905.13.2.2]	{314.16B2}
☐ Support fittings count as 1 conductor for each fitting type based on largest conductor in box_____	[3905.13.2.3]	{314.16B3}
☐ Count devices as 2 conductors based upon the connected wire size_____	[3905.13.2.4]	{314.16B4}
☐ All EGCs count only as 1 based on largest ___	[3905.13.2.5]	{314.16B5}

FIG. 20 Device Support | Fixture Support FIG. 21

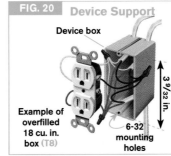

Device box

Example of overfilled 18 cu. in. box (T8)

6-32 mounting holes

3 9/32 in.

Fixture mud ring 8-32 mounting holes

Octagonal box 8-32 mounting holes

(Paddle fans req. 10-32 screws on L&L fan box)

TABLE 8	BOX FILL EXAMPLE, F20		
Item	Size	#	Total
#12 conductors exiting box	2.25	6	13.50
Largest grounding conductor—count only 1	2.25	1	2.25
Devices—2× connected conductor size	4.50	1	4.50
Internal clamps—1 based on largest wire present	2.25	1	2.25
	TOTAL		22.5

3 12/2+G Romex + device overfills 18 cu. in. box.

GFCIs—GROUND-FAULT CIRCUIT INTERRUPTERS

A ground fault occurs when current leaks out of its normal path and finds a path back to the utility transformer through conductors that are not supposed to carry current. An example of such an abnormal path could include a human body. Ironically, even though the earth is not a sufficiently good conductor to provide a fault path that would trip a breaker, it is a good enough conductor to carry the low levels of current that can cause electrocution. GFCIs respond to very low levels of current imbalance in a circuit, such as those that occur when current returns through a person. GFCIs are designed to limit the duration of leaking current to safe levels.

How does a GFCI work its magic? In F22, equal currents are flowing to and from the load. When any electrical current flows, it generates a magnetic field. The magnetic fields generated by the flow of electrons in these two conductors are of opposite polarity (north and south, leaving and returning). The forces are equal and opposite, and their magnetic fields cancel each other. The circuit passes through a coil of wire inside the GFCI, and the GFCI accounts for the electrons on each conductor. As long as the currents are balanced, GFCI allows current on the circuit.

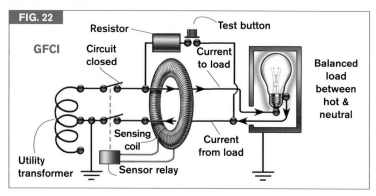

FIG. 22

During a ground fault, such as the flow of current through a person to something that is grounded, the circuit becomes unbalanced F23. Because the circuit is unbalanced, it produces a magnetic field that induces a small voltage on the sensing coil. The resulting current on the sensing coil signals the relay mechanism which opens the circuit..

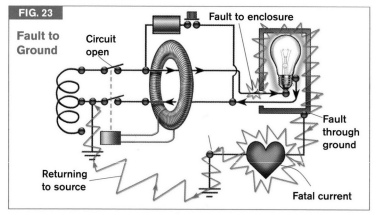

FIG. 23

A GFCI also detects improper connections of the neutral (grounded conductor) to ground. A second "injector" coil F24 surrounds the monitored circuit and induces a small current. Should the neutral have a downstream connection to ground, current will escape outside the circuit, and the sensor coil circuit will be activated as described above.

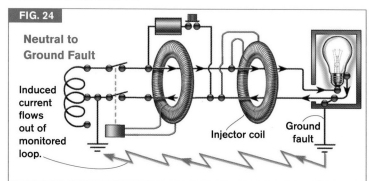

FIG. 24

GFCIs take more space inside a box than a conventional receptacle. When adding GFCIs to old houses with shallow boxes, it might be necessary to first add an extension box, as in F25.

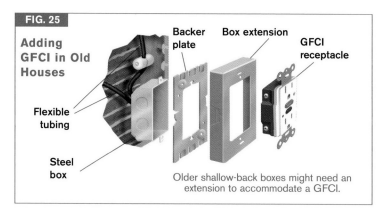

FIG. 25

Older shallow-back boxes might need an extension to accommodate a GFCI.

A GFCI will operate properly without an equipment ground. The receptacle should be labeled "no equipment ground" & any downstream protected receptacles should also have that label as well as a label stating that they are GFCI protected. Labels are not required for properly grounded GFCI-protected receptacles.

A GFCI receptacle can provide protection for other receptacles downstream on the circuit. GFCI protection can be provided by GFCI breakers or GFCI receptacles F26.

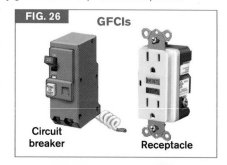

FIG. 26 GFCIs

Circuit breaker   Receptacle

Receptacle Outlets – General Purpose F27&28

	09 IRC	11 NEC
☐ Walls ≥2ft wide req receptacle	[3901.2.2]	{210.52A2}
☐ Partitions & bar-type counters count as walls F30	[3901.2.2]	{210.52A2}
☐ Doorways & fireplaces not counted as walls	[3901.2.2]	{210.52A2}
☐ Receptacle req'd within 6ft measured horizontally of any point along floor line	[3901.2.1]	{210.52A1}
☐ Receptacle req'd for hallways ≥10ft in length F28	[3901.10]	{210.52H}
☐ Receptacles that are part of electric baseboard heaters OK as req'd outlets	[3901.1]	{210.52}
☐ Receptacles >5½ft high not OK as req'd outlets	[3901.1]	{210.52}
☐ Floor receptacles >18in from wall not OK as req'd outlets	[3901.2.3]	{210.52A3}
☐ Switched receptacles installed as req'd lighting do not count as part of req'd receptacle outlets unless "half hot" [3901.1][26]		{210.52}
☐ Receptacles req'd each wall ≥3ft in foyers >60sq. ft [n/a]		{210.52I}[27]
☐ Garages & unfinished basements req min 1 receptacle in addition to any for specific equipment	[3901.9]	{210.52G}

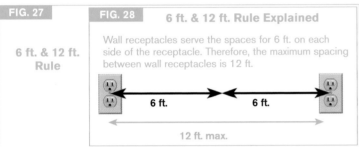

FIG. 27 — 6 ft. & 12 ft. Rule

FIG. 28 — 6 ft. & 12 ft. Rule Explained

Wall receptacles serve the spaces for 6 ft. on each side of the receptacle. Therefore, the maximum spacing between wall receptacles is 12 ft.

6 ft. | 6 ft.

12 ft. max.

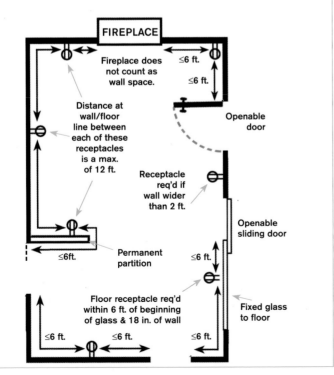

FIREPLACE

Fireplace does not count as wall space.

≤6 ft.

≤6 ft.

Distance at wall/floor line between each of these receptacles is a max. of 12 ft.

Openable door

Receptacle req'd if wall wider than 2 ft.

Openable sliding door

Permanent partition

≤6ft.

≤6 ft.

Fixed glass to floor

Floor receptacle req'd within 6 ft. of beginning of glass & 18 in. of wall

≤6 ft. | ≤6 ft. | ≤6 ft.

Bathrooms

	09 IRC	11 NEC
☐ Receptacle req'd on wall or partition within 3ft of each basin or in side or face of cabinet ≤12in below countertop	[3901.6]	{210.52D}
☐ No face-up outlets on vanity countertop	[3901.6]	{406.4E}
☐ Listed countertop-mounted receptacles OK	[n/a]	{210.52D}[28]
☐ No receptacles within or directly over tub or shower	[4002.11]	{406.9C}
☐ Separate 20A circuit for bath receptacles only OR	[3703.4]	{210.11C3}
Dedicated 20A circuit to each bathroom	[3703.4X]	{210.11C3X}
☐ Max rating of fixed space heater on general lighting circuit 15A circuit: 900W; 20A circuit: 1,200W	[3702.5]	{210.23A2}

Laundry

	09 IRC	11 NEC
☐ Min 1 20A circuit for laundry receptacles	[3703.3]	{210.11C2}
☐ No other outlets on laundry receptacle circuit	[3703.3]	{210.11C2}
☐ Receptacle within 6ft of intended appliance location	[3901.5]	{210.50C}
☐ Electric dryer min 30A circuit (10AWG Cu, 8AWG Al)	[T3704.2(1)]	{220.54}
☐ Electric dryer req's 4-conductor branch circuit EXC	[3908.7]	{250.140}
• Existing 3-wire circuits allowed to remain in use	[n/a]	{250.140X}

Outdoors

	09 IRC	11 NEC
☐ Receptacle accessible from grade req'd at front & rear of dwelling max 6½ft above grade	[3901.7]	{210.52E1}
☐ Receptacle req'd for balconies w/ interior access (balconies <20sq ft exempt in 08 NEC)	[3901.7][29]	{210.52E3}
☐ Receptacles in damp or wet locations req'd to be listed weather-resistant type	[4002.8][30]	{406.9A&B}
☐ Outdoor damp location receptacle (e.g., protected porch) req's weatherproof cover F29	[4002.8]	{406.9A}
☐ Wet location 15A & 20A receptacles req in-use covers F29	[4002.9]	{406.9B1}

FIG. 29

Outdoor Covers

Switch cover

In-use cover

Lighting Outlets (see p.19 for Switches)

	09 IRC	11 NEC
☐ Wall-switch controlled lighting outlets req'd in all habitable rooms & bathrooms	[3903.2]	{210.70A1}
☐ Habitable room lighting outlets may be switched receptacle except in kitchen & bathroom	[3903.2X1]	{210.70A1X1}
☐ Occupancy-sensor wall switches w/ manual override feature OK	[3903.2X2]	{210.70A1X2}
☐ Wall-switch controlled lighting outlets req'd in hallways, stairways, attached garages, & detached garages w/ power	[3903.3]	{210.70A2}
☐ Min 1 wall-switched lighting outlet in garage	[3903.3]	{210.70A2a}
☐ Lighting outlet req'd on exterior side grade level doors	[3903.3]	{210.70A2b}
☐ Lighting outlet req'd at garage egress doors	[3903.3]	{210.70A2b}
☐ Lighting outlet not req'd at garage vehicle doors	[3903.3]	{210.70A2b}

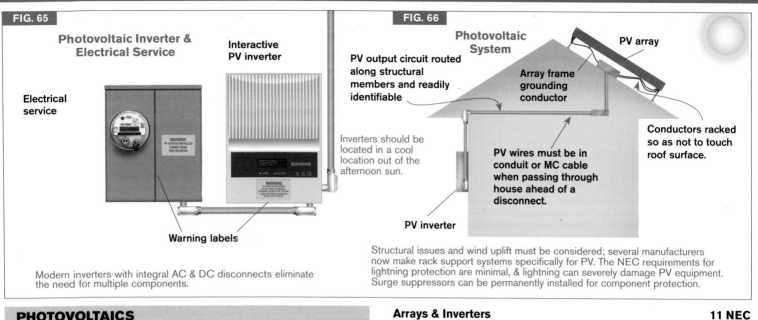

FIG. 65

Photovoltaic Inverter & Electrical Service

Electrical service

Interactive PV inverter

Warning labels

Inverters should be located in a cool location out of the afternoon sun.

Modern inverters with integral AC & DC disconnects eliminate the need for multiple components.

FIG. 66

Photovoltaic System

PV output circuit routed along structural members and readily identifiable

PV array

Array frame grounding conductor

Conductors racked so as not to touch roof surface.

PV wires must be in conduit or MC cable when passing through house ahead of a disconnect.

PV inverter

Structural issues and wind uplift must be considered; several manufacturers now make rack support systems specifically for PV. The NEC requirements for lightning protection are minimal, & lightning can severely damage PV equipment. Surge suppressors can be permanently installed for component protection.

PHOTOVOLTAICS

In most states, the utility will rebate a portion of the cost of a PV system. Time-of-use and net metering can reduce or eliminate monthly utility costs. The quality and efficiency of PV equipment have improved greatly in the last few years. What once required numerous separate components is often integrated into a single piece of equipment. Contact the utility and building department before beginning any project involving renewable energy sources.

Definitions

Array: An assembly of panels that forms the power-producing unit **F66.**

Combiner: The location where parallel PV source circuits are connected to create a PV output circuit.

Hybrid system: A system with multiple power sources (not including the utility or batteries). An example would be a system with a generator and a PV source.

Interactive system: A solar PV system that operates in parallel to the utility.

Inverter: Equipment that converts the DC current & voltage of a PV output circuit to an AC waveform **F65.**

Inverter output circuit: The AC conductors from an inverter to an AC panelboard or service **F65.**

Module: A group of PV cells connected together and encapsulated in an environmentally protective laminate—usually tempered glass—to generate DC power when exposed to the sun.

Panel: A group of modules preassembled onto a common frame and designed to be field installed.

PV output circuit: Conductors between the PV source circuits and the inverter. **F66**

PV source circuits: Circuits between modules & circuits from modules to the common connection points (combiners) of the DC system.

Stand-alone system: Solar PV system supplying power independent of the utility.

General   **11 NEC**

☐ Inverters, modules, panels, source circuit combiners L&L for PV ____ {690.4D}

☐ PV req'd to be installed only by qualified persons _____ {690.4E}[45]

☐ Max voltage = sum of rated open-circuit voltage of series-connected modules times correction factors for cold temp **F67** _____ {690.7A}

☐ All power sources req disconnects _____ {690.15}

☐ DC disconnect req'd for ungrounded conductors **F65** _____ {690.13}

☐ PV output circuits req in-sight disconnect_____ {690.16B}

☐ Disconnect for ungrounded conductors must be readily accessible switch or breaker w/ no exposed live parts **F65** _____ {690.17}

☐ Warning req'd at DC disconnect if all terminals hot while open **F65** {690.17}

☐ Rated max currents & voltages labeled on DC disconnect _____ {690.53}

☐ No disconnect on grounded conductor if it would be left energized {690.13}

☐ PV disconnecting means req'd to be on outside or inside nearest point of entrance of conductors EXC_____ {690.14(C1}

  • Source circuits through interior OK in metal conduit **F66**_____ {690.31E}

☐ AC disconnects energized from 2 directions req warning label **F65** _{690.17}

☐ Backfed breakers in load center req'd to be secured in place ____ {609.10E}

Arrays & Inverters   **11 NEC**

☐ Req'd markings on modules: polarity, max OCPD rating for module protection, open-circuit voltage, operating voltage, max system voltage, operating current, short-circuit current, & max power_____ {690.51}

☐ PV circuits may not share raceways w/ non-PV systems EXC _____ {690.4B}

  • OK w/ barriers, tagging, & grouping _____ {690.4B}

☐ DC ground-fault protection (DC GFP) req'd_____ {690.5}

☐ Inverter listed as interactive if used in interactive system_____ {690.60}

☐ DC arc-fault protection req'd systems >80V _____ {690.11}[46]

☐ Interactive systems to automatically disconnect in grid outage EXC {690.61}

  • OK to feed subpanel isolated from service by transfer switch ____ {690.61}

Grounding   **11 NEC**

☐ Module frames & all metal parts must be grounded _____ {690.43A}

☐ Size EGCs of PV output circuit per **T6** & min 14AWG _____ {690.45}

☐ EGCs must be run in same raceway as PV array circuit conductors {690.43F}

☐ Bond ground-mounted array structures_____ {690.43C}

☐ DC 2-wire system >50V must have grounded conductor _____ {690.41}

☐ Same conductor can perform DC grounding, AC grounding, & bonding between AC & DC systems **F65,66** _____ {690.47C3}

☐ When grounded conductor bonded to EGC internal within DC GFP device, bond not to be duplicated w/ an external connection _____ {690.42X}

Overcurrent Protection & Wiring   **11 NEC**

☐ Single OCPD OK for series-connected string _____ {690.9E}

☐ Sum of PV & main breakers not >120% of panel rating _____ {705.12D2}

☐ Source circuit currents = 125% × sum of parallel-circuit currents {690.8A1}

☐ Locate PV breaker opposite end of bus from main or feeder input ___ {705.12D7}

☐ Apply label warning against moving PV breaker _____ {705.12D7}

☐ Size conductors for 125% of max PV source short-circuit currents {690.8B1}

☐ Max allowable voltage in SFD 600V _____ {690.7C}

☐ Consider high ambient temps (use 90°C wire)_____ {690.31}

☐ No multiwire or 240V circuits in panels w/ 120V supply_____ {690.10C}

☐ Single conductor cables type USE or L&L as PV wire in exposed outdoor source circuits (behind modules) _____ {690.31B}

FIG. 67 **Voltage Correction Factors {NEC T690.7}**

Degrees Farenheit

	−40	−4	32	68
Multiply by this amount	1.25			
	1.18			
	1.10			
	1.02			
	−40	−20	0	20

Degrees Centigrade

Residential GFCI Protection   09 IRC   11 NEC

GFCI protection is required for 15A & 20A receptacles in the following locations. It is not required for 240V receptacles or 120V-30A receptacles.

- ☐ GFCI devices req'd to be in readily accessible locations _[n/a]   {210.8A}[19]
- ☐ All bathroom receptacles_____ [3902.1]   {210.8A1}
- ☐ All garage & accessory building receptacles_____ [3902.2][20] {210.8A2}
- ☐ All receptacles in unfinished basements EXC_____ [3902.5] {210.8A5}
  - • Permanently installed fire or burglar alarm system _ [3902.5X3] [210.8A5X]

The 2005 NEC and 2006 IRC had exceptions for receptacles in garages and unfinished basements when those receptacles served appliances that are not easily moved, such as freezers. Those exceptions have been removed.

- ☐ All outdoor receptacles EXC _____ [3902.3] {210.8A3}
  - • GFPE circuit dedicated to non-readily accessible receptacles for snow-melting or deicing equipment _____ [3902.3X] [210.8A3X]
- ☐ All receptacles in crawlspaces at or below grade level _[3902.4] {210.8A4}
- ☐ All receptacles serving kitchen counters **F30**_____ [3902.6] {210.8A6}
- ☐ Receptacles within 6ft of all non-kitchen sinks _____ [3902.7][21] [210.8A7][22]

Pools, Spas, Whirlpool Tubs, & Boathouses   09 IRC   11 NEC

- ☐ Receptacles ≤20ft of pools & outdoor hot tubs _____ [4203.1.3] {680.22A4}
- ☐ Distance does not apply to cords that would have to pass through a window or door_____ [4203.1] {680.22A5}
- ☐ Receptacles for 120V or 240V pool pump motors regardless of distance from pool_____ [4203.1.3] {680.21C}
- ☐ Receptacles providing power to indoor spas or hot tubs __ [n/a] {680.43A3}
- ☐ Receptacles ≤10ft of indoor spas or hot tubs_____ [4203.1.5] {680.43A2}
- ☐ Pool cover motor & controller _____[4206.11] {680.27B2}
- ☐ Hydromassage (whirlpool) tubs _____ [4209.1] {680.71}
- ☐ Underwater pool lights >15V _____ [4206.4] {680.23A3}
- ☐ Luminaires & lighting outlets <10ft horizontally from outdoor pool or spa edge unless >5ft vertically above water_____ [4203.4.5] {680.22B4}
- ☐ Existing luminaires allowed <5ft horizontal if >5ft vertical above water & GFCI protected_____ [4203.4.3] {680.22B3}
- ☐ Outlets supplying self-contained packaged spa/hot tub or field-assembled w/ heating <50A EXC _____ [4208.1] {680.44}
  - • Outlets supplying listed units w/ integral GFCIs _____ [4208.1] {680.44A}
- ☐ Receptacles in boathouses_____ [3902.8] {210.8A8}
- ☐ 120V or 240V boat hoists_____ [3902.9][23] {210.8C}

UL 943–the standard of safety for GFCIs–was revised in 2003, requiring GFCIs to have greater resistance to corrosion and surges. GFCIs have become more reliable and do not have the problems of "nuisance tripping" that characterized these devices in the earlier stages of their development. Thanks to their increased reliability, it is no longer necessary to have the numerous exceptions that once existed for GFCIs associated with motor loads. The new standard included a line-load reversal test requiring that the receptacle not be capable of resetting if it is miswired, and a 2006 revision requires that there be no power to the face of a miswired receptacle. The contacts on newer GFCIs ensure proper resetting and prevent some miswiring that could appear from manipulation of the controls on the older GFCIs. In addition, manufacturer's installation instructions for GFCIs are now standardized for consistency. These instructions require specific methods for checking GFCI operation after installation to ensure that devices are properly wired and that they be tested on a regular basis for the life of the GFCI. As a result, these proven life savers have become more reliable than ever.

BRANCH CIRCUITS & OUTLETS

Branch circuits are the permanent wiring between the final overcurrent protective devices (fuses or breakers) and the lighting or receptacle outlets from which electrical equipment derives power. During rough-in of branch circuit wiring, care should be taken to ensure they are an adequate size for the load. Circuits for continuous loads and items such as water heaters or space heaters that are treated as continuous loads must be sized to 125% of the load. There must be sufficient outlets for the needs of the occupants. An insufficient number of outlets could lead to the dangerous substitution of extension cords in place of permanent wiring. During rough-in, boxes are placed in the locations required for receptacle and lighting outlets, cables are run, and equipment grounds are connected.

Circuit Sizes, Number, & Load Limitations   09 IRC   11 NEC

- ☐ Rule of thumb: min 1 general-purpose circuit per 500sq ft _____ [3704.4] {220.12}
- ☐ Load not to exceed rating of branch circuit_____ [3701.2] {220.18}
- ☐ Min circuit size 125% of the continuous load + 100% of the noncontinous load _____ [3701.2] {210.19A}
- ☐ Continuous load = max current for 3 hours or more____[3501] {100}
- ☐ Min size for branch circuit wiring 14AWG _____ [3702.13] {210.19A4}
- ☐ Branch circuit ratings for other than individual circuits must be 15A, 20A, 30A, 40A, or 50A _____ [3702.2] {210.3}
- ☐ Single piece of cord-&-plug-connected equipment not permanently fastened in place cannot exceed 80% of 15A or 20A branch circuit _____ [3702.3] {210.23A1}
- ☐ Max single cord-&-plug-connected load on multi-receptacle circuit not to exceed 80% of circuit rating_____[n/a] {210.21B2}

Receptacles   09 IRC   11 NEC

- ☐ All receptacles on 15A & 20A circuits grounding type [4002.2] {406.4A}
- ☐ Receptacles for direct Al connection marked "CO/ALR" [4002.3] {406.3C}
- ☐ All req'd receptacles listed tamper-resistant type EXC [4002.14][24] {406.12}
  - • Receptacles located >5½ ft above floor _____ [n/a] {406.12X}[25]
  - • Receptacles that are part of a luminaire _____ [n/a] {406.12X}[25]
  - • Single receptacles within dedicated space for an appliance not easily moved or duplex receptacle for 2 such appliances _[n/a] {406.12X}[25]
  - • Replacement nongrounding receptacles (see **p. 29**)_____[n/a] {406.12X}[25]
- ☐ Single receptacles rated not less than branch circuit[4002.1.1] {210.21B1}
- ☐ Multiple receptacles on branch circuit per **T9**_____ [4002.1.2] {210.21B3}

For the purposes of these rules, a duplex receptacle is two receptacles, not a "single" receptacle.

TABLE 9	RECEPTACLE RATINGS FOR MULTIPLE RECEPTACLES ON 1 CIRCUIT [4002.1.2] & {210.21.B3}
Circuit Rating (amperes)	**Receptacle Rating (amperes)**
15	not over 15
20	15 or 20
30	30
40	40 or 50
50	50

Receptacle Locations – General   09 IRC   11 NEC

- ☐ Receptacles for specific appliances (laundry, garage door opener) within 6ft of appliance location _____ [3901.5] {210.50C}
- ☐ Flexible cords not OK as fixed or concealed wiring __ [3909.1] {400.8}

KITCHENS

A minimum of two 20A small-appliance branch circuits are required for portable appliances that are used in kitchens and dining areas. These circuits are in addition to those that supply lighting or permanently installed appliances. Portable kitchen appliances have short cords so they are not as likely to be run across cooktops or sinks or to hang down in the reach of children. A receptacle is needed to serve every countertop 1 ft. or more in width.

Branch Circuits
	09 IRC	11 NEC
☐ Min 2 20A small-appliance circuits req'd	[3703.2]	{210.11C1}
☐ Small-appliance circuits must serve refrigerator & all countertop & exposed wall receptacles in kitchen, dining room, & pantry EXC	[3703.2]	{210.52B1}
• Refrigerator OK on individual branch circuit ≥15A	[3703.2X]	{210.52B1X2}
☐ Switched receptacle for dining room light OK on non-small-appliance circuit	[3901.3X1]	{210.52B1X1}
☐ No other outlets (including lights) on small appliance branch circuits EXC	[3901.3.1]	{210.52B2}
• Receptacles for clock or gas range ignition OK	[3901.3.1X]	{210.52B2X}
☐ Dishwasher & disposer req separate circuits if combined rating exceeds branch circuit rating	[3701.2]	{210.19A1}
☐ Circuits for ranges ≥8.75kW min 40A 240V	[3702.9.1]	{210.19A3}

Receptacles for Countertop Spaces
	09 IRC	11 NEC
☐ Receptacles req'd for wall counter spaces ≥12in wide	[3901.4.1]	{210.52C1}
☐ Countertop spaces separated by sinks or ranges considered separate countertop spaces F30	[3901.4.4]	{210.52C4}
☐ Spacing so no point >24in from receptacle F31	[3901.4.1]	{210.52C1}
☐ Area behind sink or range considered countertop space if ≥12in to wall F32 or ≥18in to corner F33	[3901.4.1X]	{210.52C1X}
☐ Max 20in above countertop	[3901.4.5]	{210.52C5}
☐ No face-up countertop receptacles EXC	[3901.4.5]	{406.4E}
• L&L assemblies installed AMI	[n/a]	{210.52C5}[31]
☐ Peninsulas req receptacle if long dimension ≥24in & short dimension >12in, measured from connecting edge F30	[3901.4.3]	{210.52C3}
☐ Island & peninsula countertop spaces min 1 receptacle per space—no 24in rule F30	[3901.4.2&3]	{210.52C2&3}
☐ Sink or range w/ <12in behind divides counters into separate spaces for above rule	[3901.4][32]	{210.52C4}
☐ Island & peninsula receptacles OK ≤12in below counter overhanging ≤ & no means of installing receptacle in an overhead cabinet F30	[3901.4.5X]	{210.52C5X}
☐ GFCI protection for all receptacles serving countertops	[3902.6]	{210.8A6}

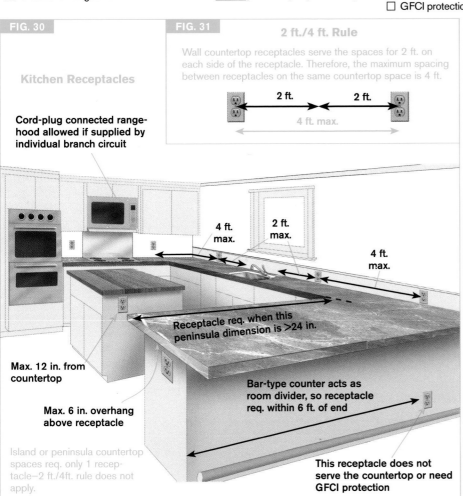

FIG. 30 Kitchen Receptacles

Cord-plug connected range-hood allowed if supplied by individual branch circuit

4 ft. max.

2 ft. max.

4 ft. max.

Receptacle req. when this peninsula dimension is >24 in.

Max. 12 in. from countertop

Max. 6 in. overhang above receptacle

Island or peninsula countertop spaces req. only 1 receptacle—2 ft./4ft. rule does not apply.

Bar-type counter acts as room divider, so receptacle req. within 6 ft. of end

This receptacle does not serve the countertop or need GFCI protection

FIG. 31 2 ft./4 ft. Rule

Wall countertop receptacles serve the spaces for 2 ft. on each side of the receptacle. Therefore, the maximum spacing between receptacles on the same countertop space is 4 ft.

2 ft. — 2 ft.

4 ft. max.

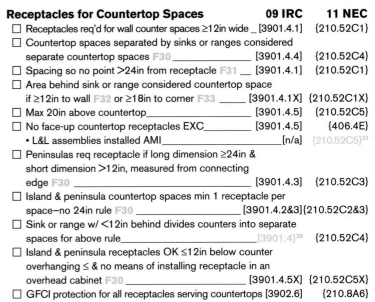

FIG. 32 Extended Range or Sink

If X ≥ 12 in., countertops not considered separate spaces & the 2 ft./4 ft. rule applies to the entire countertop.

X <12 in.: measure from here

X <12 in.: measure from here

X

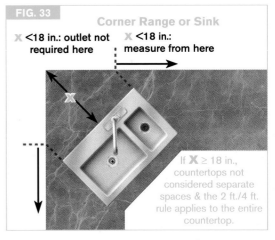

FIG. 33 Corner Range or Sink

X <18 in.: outlet not required here

X <18 in.: measure from here

X

If X ≥ 18 in., countertops not considered separate spaces & the 2 ft./4 ft. rule applies to the entire countertop.

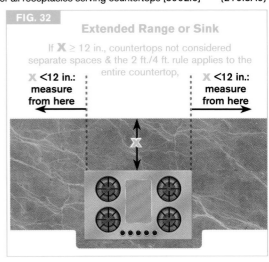

SWITCHES

Switches	09 IRC	11 NEC
☐ All switching in ungrounded conductors **F34,35** __ [4001.8&9]		{404.2A&B}
☐ Provide neutral in switchbox EXC _____ [n/a]		{404.2C}[33]
• In raceway w/ sufficient room to add neutral _____ [n/a]		{404.2CX}[34]
• Where switch not enclosed by building finishes_____ [n/a]		{404.2CX}[34]
☐ Snap switches & dimmers req grounding EXC___ [4001.11.1]		{404.9B}
• Replacements where no grounding means present OK w/ plastic faceplate or GFCI protection _____ [4001.11.1X]		{404.9BX}
☐ Grounding OK by screws to grounded metal box [4001.11.1]		{404.9B1}
☐ Metal faceplates must be grounded to switch ___ [4001.11.1]		{404.9B}
☐ Faceplate must completely cover wall opening_____ [4001.11]		{404.9A}
☐ Switch at each entrance of stairs w/ ≥6 risers _____ [3903.3]		{210.70A2c}
☐ Dimmers only for incandescent lights not receptacles [4001.12]		{404.14E}
☐ Current-carrying conductors of circuit grouped **F34** _ [3406.7]		{300.3B}
☐ Re-identify ungrounded white or gray wires **F34** ___ [3407.3X]		{200.7C}
☐ "CO/ALR" switch req'd if direct Al wire connection__ [4001.2]		{404.14C}

FIG. 34

3-Way Switch

3-way switching takes place from a common terminal to one or the other "travelers."

This traditional method of running power first to the luminaire & then to 3-way switches with a common wire & 2 travelers is no longer allowed unless the cable also contains a neutral conductor of the circuit. (4 conductor +G cable would be OK).

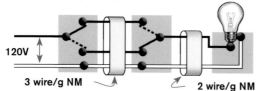

Common, Travelers, Switch up

120V

2 wire/g NM   3 wire/g NM

Acceptable 3-way switching w/ neutral in each switch enclosure

Switch down

120V

3 wire/g NM   2 wire/g NM

(Equipment ground not shown but required for any new installation)

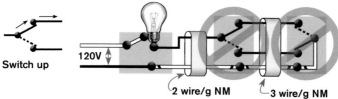

FIG. 35

4-Way Switch

A 4-way switch is a double-pole double-throw switch. Any number can be placed between the two 3-ways.

4-way interrupts travelers

120V

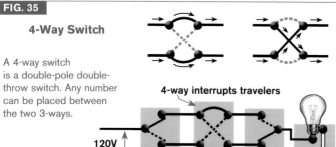

LIGHTING

Lighting outlets and luminaires must be installed with no exposed live parts that could pose a shock hazard. The heating effect of luminaires must be considered, especially around thermal insulation. Lights rated "type IC" are suitable for insulated ceilings. See **p.17** for required locations.

General	09 IRC	11 NEC
☐ All luminaires & lampholders listed _____ [3403.3]		{410.6}
☐ Exposed metal parts grounded EXC _____ [4003.3]		{410.42A}
• Incidental metal parts such as mounting screws ___ [4003.3]		{410.42A}
☐ Wet location luminaires L&L for wet location_____ [4003.8]		{410.10A}
☐ Damp location luminaires L&L for damp or wet location [4003.8]		{410.10A}
☐ Screw shells for lampholders only–no adapters _____ [4003.4]		{410.90}

Recessed Lights	09 IRC	11 NEC
☐ Non-Type IC min ½in from combustibles _____ [4004.8]		{410.116A1}
☐ Non-Type IC min 3in from insulation_____ [4004.9]		{410.116B}
☐ Type IC OK in contact w/ combustible material _____ [4004.8]		{410.116A2}
☐ Type IC OK in contact w/ insulation_____ [4004.9]		{410.116B}
☐ Luminaires that req >60°C wire must be marked _____ [n/a]		{410.74}
☐ Connect proper temp-rated wire to luminaire _____ [n/a]		{410.117A}
☐ Tap conductors to 60°C wire min 18in max 6ft **F36**_____ [n/a]		{410.117C}

FIG. 36

Recessed Lighting with Old Wiring

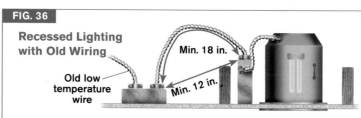

Old low temperature wire

Min. 18 in.

Min. 12 in.

Closet Lights **F37**	09 IRC	11 NEC
☐ Incandescent bulbs req'd to be fully enclosed _____ [4003.12]		{410.16A1}
☐ Partially enclosed incandescent bulbs prohibited___ [4003.12]		{410.16B}
☐ Surface-mounted only on ceiling or wall above door [4003.12]		{410.16C1&2}
☐ Surface incandescents min 12in from storage _____ [4003.12]		{410.16C1}
☐ Surface fluorescents min 6in from storage _____ [4003.12]		{410.16C2}
☐ Recessed (wall or ceiling) min 6in from storage ____ [4003.12]		{410.16C3&4}
☐ Surface fluorescent or LED (light-emitting diode) OK in storage area if listed for same_____ [4003.12][35]		{410.16C5}

FIG. 37

Closet Lights

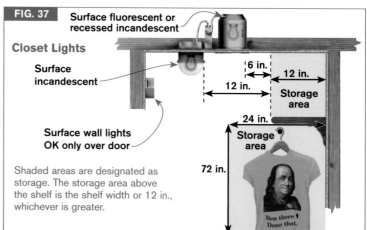

Surface fluorescent or recessed incandescent

Surface incandescent

Surface wall lights OK only over door

6 in.   12 in.   12 in.

Storage area

24 in.

Storage area

72 in.

Shaded areas are designated as storage. The storage area above the shelf is the shelf width or 12 in., whichever is greater.

Ben there t Done that.

Track Lighting

	09 IRC	11 NEC
☐ Branch circuit rating ≤track rating	[4005.1]	{410.151B}
☐ Connected load ≤track rating	[4005.3]	{410.151B}
☐ No track concealed, extended through walls or partitions, or in damp or wet locations	[4005.4]	{410.151C}
☐ Track must be securely fastened	[4005.5]	{410.154}
☐ Track must be grounded	[4005.6]	{410.155B}

Tub & Shower Areas F38

	09 IRC	11 NEC
☐ No cord-connected or pendant luminaires, lighting track, or ceiling-suspended paddle fans first 8ft above tub rim or shower threshold & for zone extending 3ft outside	[4003.10]	{410.10D}
☐ Luminaires directly above tub & shower listed for damp locations (or wet locations if subject to shower spray)	[4003.10]	{410.10D}

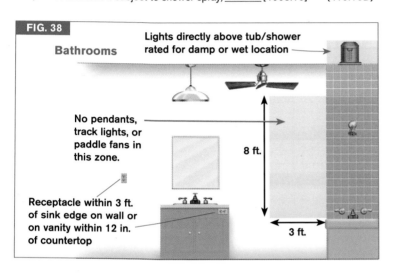

FIG. 38

Bathrooms

Lights directly above tub/shower rated for damp or wet location

No pendants, track lights, or paddle fans in this zone.

8 ft.

3 ft.

Receptacle within 3 ft. of sink edge on wall or on vanity within 12 in. of countertop

APPLIANCES

The term *appliances* is a generic term for standardized manufactured equipment that utilizes electricity (other than lighting). Whether portable or permanent, all appliances require a means of disconnecting the power source so the appliance can be safely serviced or replaced. The codes provide general rules for disconnecting appliances as well as specific rules for common built-in (fixed in place) appliances.

Disconnecting Devices

	09 IRC	11 NEC
☐ All appliances req disconnecting means	[4101.5]	{422.30}
☐ Cord-connected appliances req attachment plug	[3909.4]	{400.7B}
☐ Accessible attachment plug OK as disconnect	[T4101.5]	{422.33A}
☐ Additional disconnect req'd if plug not accessible	[T4101.5]	{422.33A}
☐ Breaker alone OK for appliances <300VA or 1/8hp	[T4101.5]	{422.31A}
☐ In-sight switch or breaker req'd if ≥300VA or 1/8hp, or lockable breaker OK when not in sight F39	[T4101.5]	{422.31B&C}
☐ Breaker lockouts req permanent hasp F39	[T4101.5]	{422.31B}
☐ Unit switch opening all ungrounded conductors OK	[T4101.5]	{422.34}

FIG. 39

Breaker Lockout

Hydromassage Tub (Whirlpool Bathtub)

	09 IRC	11 NEC
☐ Readily-accessible GFCI protection req'd	[4209.1]	{680.71}
☐ Individual branch circuit req'd	[4209.1][36]	{680.71}
☐ Electrical equipment (pump motor) must be accessible	[4209.3]	{680.73}
☐ Disconnecting means req'd in sight of motor	[T4101.5]	{430.102B}
☐ Bond metal parts in contact w/ circulating water	[4209.4]	{680.74}
☐ Bonding conductor min solid 8AWG Cu F40	[4209.4]	{680.74}
☐ Bond metal piping system to motor lug EXC F40	[4209.4][37]	{680.74}
• Double-insulated motor	[4209.4]	{680.74}
☐ Bonding conductor need not connect to panelboards	[4209.4]	{680.74}

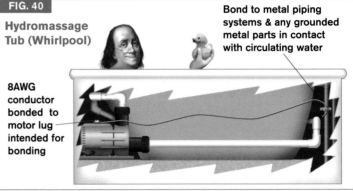

FIG. 40

Hydromassage Tub (Whirlpool)

Bond to metal piping systems & any grounded metal parts in contact with circulating water

8AWG conductor bonded to motor lug intended for bonding

Kitchens

	09 IRC	11 NEC
☐ Cords only OK on appliances listed for same	[4101.3]	{422.16A}
☐ Garbage disposer cord min 18in max 36in	[T4101.3]	{422.16B1}
☐ Dishwasher or trash compactor cord min 3ft max 4ft measured from back	[T4001.3]	{422.16B2}
☐ Dishwasher & compactor receptacles in same space as appliance or in adjacent space	[n/a]	{422.16B2}
☐ Range hoods can be cord & plug connected if L&L for cord & on individual branch circuit	[4101.3]	{422.16B4}
☐ Range hood cords min 18in, max 36in	[T4101.3]	{422.16B4}
☐ Cord & plug ovens & cooking units OK if L&L	[4101.3]	{422.16B3}

Central Furnace

	09 IRC	11 NEC
☐ In-sight disconnect req'd	[T4101.5]	{422.32}

Refer to manu instructions for possible supplemental OCPD reqs F41

☐ Lighting outlet switched at entry to equipment space	[3903.4]	{210.70A3}
☐ Central furnace must be on individual circuit EXC	[3703.1]	{422.12}
• Associated equipment (electrostatic filters, pumps, etc)	[3703.1]	{422.12X1}
☐ 120V receptacle req'd within 25ft on same elevation	[3901.11]	{210.63}

FIG. 41

SSU Switch

A fused disconnect provides supplementary overcurrent protection and is sometimes a manufacturer's instruction.

An example might be a furnace requiring 15A overcurrent protection installed on a 20A circuit.

Electric Furnaces & Space Heaters

	09 IRC	11 NEC
☐ Branch circuit 125% load (heat watts + motor)	[3702.10]	{424.3B}
☐ Disconnect in sight or lockable breaker F39	[T4101.5]	{424.19}
☐ Unit switch that opens all ungrounded conductors OK as disconnect for space heater w/ no motor >1/8hp	[T4101.5]	{424.19C}

Central Vacuum
	09 IRC	11 NEC
☐ Max 80% individual branch circuit rating, 50% of multi-outlet branch circuit rating _____	[3702.3]	{210.23A}
☐ Cord must have same ampacity as branch circuit _____	[n/a]	{422.15B}
☐ Bond all non-current-carrying metal parts _____	[3908.2]	{422.15C}

Water Heater
	09 IRC	11 NEC
☐ In-sight or lockable breaker or switch OK F39 ____	[T4101.5]	{422.31B}
☐ Breaker lockout hasp req'd to remain in place w/ lock removed F39 _____	[T4101.5]	{422.31B}
☐ Bond hot, cold, & gas pipes F13 _____	[3609.7]	{250.104}

Outdoor De-icing & Snow Melting Equipment
	09 IRC	11 NEC
☐ GFPE protection req'd for de-icing equipment_____	[4101.7]	{426.28}

Some jurisdictions allow the GFPE function of an AFCI to meet this rule

Paddle Fans F42
	09 IRC	11 NEC
☐ Listed box for fan support (no standard boxes)_____	[3905.9]	{314.27C}
☐ Listed fan boxes w/o weight marking OK up to 35lb _	[3905.9]	{314.27C}
☐ Fan >35lb & <70lb, fan box L&L for suitable weight	[3905.9]	{314.27C}
☐ Independent support for fans >70lb _____	[3905.9]	{314.27C}

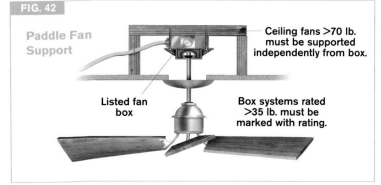

FIG. 42

Paddle Fan Support

Ceiling fans >70 lb. must be supported independently from box.

Listed fan box

Box systems rated >35 lb. must be marked with rating.

Air-Conditioning
	09 IRC	11 NEC
☐ Wiring & OCPD per nameplate of L&L equipment __	[3702.11]	{440.4B}
☐ Disconnect on or within sight of condenser F43 ___	[T4101.5]	{440.14}
☐ Disconnect not OK on compressor access panel _____	[n/a]	{440.14}
☐ Working space req'd in front of disconnect F43 ____	[3405.1]	{110.26A}
☐ Room AC plug disconnect OK if controls ≤6ft of floor____	[n/a]	{440.63}
☐ Max cord length 120V = 10ft, 240V = 6ft_____	[n/a]	{440.64}
☐ AFCI or LCDI (leakage current detection interrupter) in cord or plug for room AC units_____	[n/a]	{440.65}

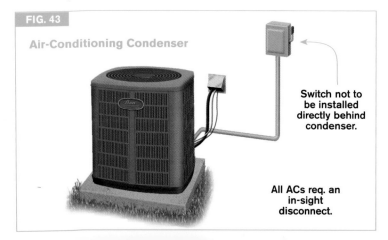

FIG. 43

Air-Conditioning Condenser

Switch not to be installed directly behind condenser.

All ACs req. an in-sight disconnect.

Smoke Alarms
	09 IRC
☐ NFPA 72 systems OK if permanent part of property _____	[314.2][38]
☐ Alarms req'd in each sleeping room & adjoining areas F44_____	[314.3]
☐ Req'd each story including basements & habitable attics F44 _____	[314.3]
☐ Interconnect so activation of 1 alarm sets off all alarms_____	[314.3]
☐ Power from building wiring & battery backup EXC_____	[314.4]
• Battery-only OK alterations or repairs w/ no access to wire path__	[314.4X2]

Carbon Monoxide Alarms
	09 IRC
☐ Req'd outside sleeping areas in dwellings w/ fuel-fired appliances or w/ attached garages F44 _____	[315.1][39]
☐ Req'd when remodeling requiring permit is performed _____	[315.2][39]
☐ Install AMI & in compliance w/ UL 2034 _____	[315.3][39]

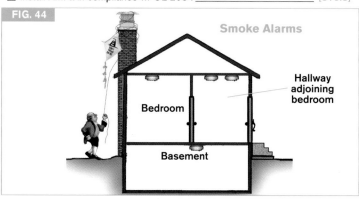

FIG. 44

Smoke Alarms

Hallway adjoining bedroom

Bedroom

Basement

T10 is a "quick reference" guide to the maximum size breaker for a given size of wire. It is an abbreviated version of the tables on the next page. Always consider if the conductors must be "derated" for ambient temperature, grouping, or the other factors on the next page. The sizes given for service entrance conductors apply only for wires with insulation types RHH, RHW, RHW-2, THHN, THHW, THW, THW-2, THWN, THWN-2, XHHW, XHHW-2, SE, USE, & USE-2.

TABLE 10	SIZING CONDUCTORS			
Fuse or Breaker	**Branch Circuits or Feeders Wire Size**		**Service Conductors Wire Size**	
	Cu	Al	Cu	Al
15	14	12	n/a	n/a
20	12	10	n/a	n/a
30	10	8	n/a	n/a
40	8	6	n/a	n/a
50	6	4	n/a	n/a
60	6	3	n/a	n/a
70	4	2	n/a	n/a
80	3	1	n/a	n/a
90	2	1/0	n/a	n/a
100	2	1/0	4	2
110	1	1/0	3	1
125	1/0	1/0	2	1/0
150	1/0	2/0	1	2/0
200	3/0	4/0	2/0	4/0
225	4/0	250kcmil	3/0	250kcmil
400	500kcmil	700kcmil	400kcmil	600kcmil

AMPACITY OF WIRE

When wire overheats, its insulation begins to break down, and we say the wire has exceeded its ampacity. Protecting conductors and equipment from overheating and insulation failure is one of the main principles of electrical safety.

General   09 IRC   11 NEC

☐ Protect conductors at their ampacity EXC _____ [3705.5]   {240.4}
  • Small conductors protected per note A in T11____[3705.5.3] {240.4D}
  • Air-conditioning protected AMI _____ [3705.5.4] {240.4G}
☐ OCPD for NM cable not to exceed 60°C ampacity _[3705.4.4]   {334.80}

Derating   09 IRC   11 NEC

☐ Apply temp-correction factor T12_____ [3705.2]   {310.15B2}
☐ Add correction for rooftop conduits per T13_____[n/a]   {310.15B3c}
☐ Derate for >3 current-carrying conductors in raceway or
  cables grouped w/o spacing >24in in length _____ [3705.3]   {310.15B3a}
☐ Derate >2 NM cables in caulked (fireblocked) hole __[3705.4.4] {334.80}
☐ Derate >2 NM cables installed w/o spacing
  in contact w/ thermal insulation _____ [3705.4.4][40] {334.80}

The first step in determining the allowable ampacity of a conductor is to look it up in T11 based on the wire size and insulation type. The most common ratings of conductor insulation are 60°C, 75°C, and 90°C. We use the 90°C column only for derating (temperature corrections), not for selection of the breaker or fuse. Conductors can be dual rated, with 75°C ratings in wet locations and 90°C ratings in dry locations, such as THWN/THHN.

Breaker and equipment terminations have a temperature rating, typically 60°C and/or 75°C. The overall ampacity of a circuit is limited by the lowest-rated device or conductor in the circuit. The final choice of breaker is, therefore, usually limited by the temperature rating of the breaker terminals; and the insulation rating is used in the derating calculations. Nonmetallic sheathed cable and SE cable as interior wiring are restricted to a 60°C rating despite containing 90°C rated conductors.

In addition to size, material, and insulation type, other factors must be considered. These are ambient temperature T12, the rate of heat dissipation into the ambient medium, and the adjacent load-carrying conductors T14. Heat dissipates more readily to free air than to water, such as found in underground conduits. Thermal insulation traps heat, as do adjacent conductors when they are grouped together.

To determine the ambient temperature correction, apply the factors of T12 to the ampacity listed in the appropriate column of T11. The heating effect of reflected sunlight must also be added to the temperature correction, per T13.

TABLE 12	AMBIENT TEMPERATURE CORRECTION [T3705.2] & {310.15B2a}			
Ambient Temp. °C	For Ambient Temps. >30°C (86°F), Multiply the Allowable Ampacities in T11 by the Following Percentages:			Ambient Temp. °F
	60°C	75°C	90°C	
31–35	0.91	0.94	0.96	87–95
36–40	0.82	0.88	0.91	96–104
41–45	0.71	0.82	0.87	105–113
46–50	0.58	0.75	0.82	114–122
51–55	0.41	0.67	0.76	123–131
56–60	–	0.58	0.71	132–140
61–70	–	0.33	0.58	141–158

This table may have little effect on post-1984 90°C-based NM-B wiring. It can be important in remodels w/ older 60°C wire.

TABLE 13	TEMPERATURE ADJUSTMENT FOR CONDUITS EXPOSED TO SUNLIGHT ABOVE ROOFTOPS {310.15B3c}	
Distance between Roof & Conduit	Temperature Added to T12	
0–1/2 in.	33°C	60°F
>1/2 in.–3 1/2 in.	22°C	40°F
>3 1/2 in.–12 in.	17°C	30°F
>12 in.	14°C	25°F

Another consideration is conductor proximity, which can trap or heat or prevent heat dissipation when conductors are grouped. When there are more than 3 current-carrying conductors in a raceway, the derating factors of T14 must be applied, in addition to any ambient temperature correction. These same derating factors also apply to a grouping of cables installed without spacing for a length of 24 in. or more and for groups >2 NM cables passing through an opening in wood framing that is fireblocked with thermal insulation, caulk, or foam and to NM cables installed without spacing and in contact with thermal insulation.

TABLE 11	WIRE AMPACITIES [T3705.1] {310.15B16}						
Size	60°C	75°C	90°C	60°C	75°C	90°C	Size
	140°F	167°F	194°F	140°F	167°F	194°F	
	INSULATION TYPES						
Cu AWG kcmil	TW UF	THHW THW THWN USE	THHN THHW THW-2 THWN2 USE-2	TW UF	XHHW USE	USE-2, XHHW-2	Al AWG kcmil
	Cu			Al			
14[A]	15	20	25	–	–	–	–
12[A]	20	25	30	15	20	25	12
10[A]	30	35	40	25	30	35	10
8	40	50	55	35	40	45	8
6	55	65	75	40	50	55	6
4	70	85	95	55	65	75	4
3	85	100	115	65	75	85	3
2	95	115	130	75	90	100	2
1	110	130	145	85	100	115	1
1/0	125	150	170	100	120	135	1/0
2/0	145	175	195	115	135	150	2/0
3/0	165	200	225	130	155	175	3/0
4/0	195	230	260	150	180	205	4/0
250	215	255	290	170	205	230	250

A. Max 30A OCPD for #10Cu, 20A for #12, & 15A for #14.
For .Al, max .25A for #10 & 20A for #12.

TABLE 14	DERATING FOR CONDUCTOR PROXIMITY [T3705.3] {310.15B3a}	
Number of Current-Carrying Wires	Ampacity Correction	
4–6	80	
7–9	70	
10–20	50	

With modern 90°C small conductors this table becomes significant when there are >9 current-carrying conductors in a conduit or cable group, or when compounded by temperature corrections. Cables installed without spacing > 2 ft. are subject to the above derating. When newer 90°C wire is connected to older 60°C wire, such as pre-1984 NM, the ampacity of the lower-rated conductors applies to the entire circuit.

CABLE SYSTEMS

Cable systems are the most common residential wiring methods. Cables contain all conductors of the circuit inside a protective outer sheath of metal or plastic. Starting with the 2005 edition, the NEC uses a parallel numbering system for rules pertaining to cables and raceways. See the common numbering system table (**T24**) on the inside back cover.

Cable Protection Indoors (NM, AC, MC, UF, SE) 09 IRC   11 NEC

☐ Bored holes & standoff clamps 1¼in setback **F56** __ [3802.1]   {300.4A&D}
☐ Protect cables w/ ¹/₁₆in steel plate {or L&L plate}
   if closer than 1¼in to framing surfaces **F45** ____ [3802.1]   {300.4A&D}
☐ Cables min 1½ in below sheet steel roof decks _____[n/a]   {300.4E}[41]
☐ Provide guard strips within 6ft of attic scuttle
   (& up to 7ft high if attic has permanent access____ [3802.2.1]   {334.23}

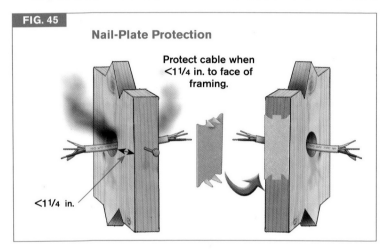

FIG. 45

Nail-Plate Protection

Protect cable when <1¼ in. to face of framing.

<1¼ in.

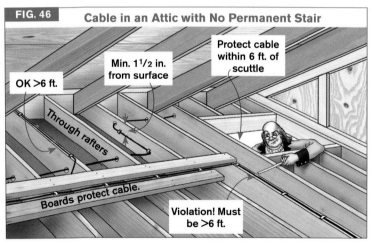

FIG. 46   **Cable in an Attic with No Permanent Stair**

Min. 1¹/₂ in. from surface

Protect cable within 6 ft. of scuttle

OK >6 ft.

Through rafters

Boards protect cable.

Violation! Must be >6 ft.

FIG. 47   **Underfloor Cable in Basement or Crawlspace**

◄ <8/3 NM cable

◄ ≥8/3 NM cable

◄ <8/3 NM cable

◄ 1×4 backing strip

NM–Nonmetallic Sheathed Cable (NM) F48   09 IRC   11 NEC

☐ OK in dry locations only_____[3801.4]   {334.12B4}
☐ Protect exposed cable from damage where necessary[3802.3.2]   {334.15B}
☐ Listed grommets for holes through metal framing____ [3802.1]   {300.4B1}
☐ OCPD selection based on 60° column **T11** _____[3705.4.4]   {334.80}
☐ Derating & temp correction based on 90° rating _____ [3705.4.4]   {334.80}
☐ Derate >2 NM cables in same caulked
   (fireblocked) hole _____[3705.4.4]   {334.80}
☐ Derate >2 NM cables installed w/o spacing in contact
   w/ thermal insulation _____ [3705.4.4][40]   {334.80}
☐ Secure to box w/ approved NM clamp EXC **F49** __ [3905.3.2] {314.17B&C}
   • Single gang (2¼×4in) plastic box stapled
   within 8in_____ [3905.3.2]   {314.17CX}
☐ Min ¼in sheathing into plastic boxes _____ [3905.3.1]   {314.17C}
☐ Secure within 12in of box & max 4½ft intervals _____ [3802.1]   {334.30}
☐ Do not overdrive staples or staple flat cable on edge [3802.1]   {334.30}
☐ Bends gradual (min 5× cable diameter) _____ [3802.5]   {334.24}
☐ Running board for small cable under joists **F47** _____ [3802.4]   {334.15C}

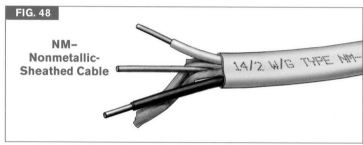

FIG. 48

NM– Nonmetallic- Sheathed Cable

14/2 W/G TYPE NM–

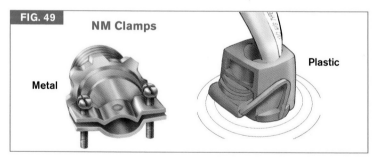

FIG. 49   NM Clamps

Metal

Plastic

AC–Armored Cable (BX™) F50   09 IRC   11 NEC

☐ Dry locations only_____[3801.4]   {320.10}
☐ Secure within 12in of box & max 4½ft intervals EXC [3802.1]   {320.30B}
   2ft where flexibility needed (motors)_____[3802.1]   {320.30D}
☐ Insulated (anti-short) bushing at terminations **F50** ___ [3802.1]   {320.40}
☐ Armor is EGC–don't bring bond wire into box **F50** __ [3908.8]   {250.118}
☐ Underside of joists–secure at each joist _____[n/a]   {320.15}

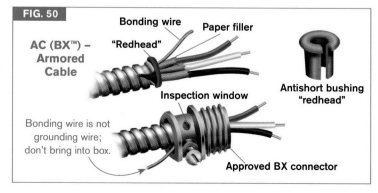

FIG. 50

AC (BX™) – Armored Cable

Bonding wire   Paper filler

"Redhead"

Inspection window

Antishort bushing "redhead"

Bonding wire is not grounding wire; don't bring into box.

Approved BX connector

UF–Underground Feeder F51

	09 IRC	11 NEC
☐ Interior installation same rules as NM _____	[3801.4]	{340.10}
☐ May be buried in earth w/ cover per T1,F52 ____	[3801.4]	{340.10}
☐ Protect where emerging from earth from 18in below grade to 8ft above F52	[3803.3]	{300.5D1}
☐ Single conductors in trench must be grouped ____	[3803.8]	{340.10}
☐ UV-resistant type OK exposed to sunlight _____	[3802.3.3]	{340.12}
☐ May not be strung through air w/o support messenger	[3802.1]	{340.12}

FIG. 51

UF – Underground Feeder Cable

UF 14/2

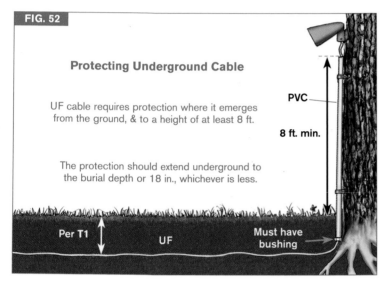

FIG. 52

Protecting Underground Cable

UF cable requires protection where it emerges from the ground, & to a height of at least 8 ft.

The protection should extend underground to the burial depth or 18 in., whichever is less.

PVC

8 ft. min.

Per T1   UF   Must have bushing

SE – Service Entrance & USE – Underground Service Entrance F53

	09 IRC	11 NEC
☐ OK as service entrance conductor (see **p. 4**) ____	[3801.4]	{338.10A}
☐ Type SE interior installation same rules as NM ___	[3802.1][42]	{338.10B4a}
☐ Type USE not OK for interior wiring _____	[3801.4]	{338.12B}
☐ SE not OK for direct burial, USE OK for direct burial	[3801.4]	[338.12B]
☐ Bare neutral OK for EGC of 240V branch circuit ___	[3801.4]	{338.10B2}
☐ Insulated neutral (type SE-R) req'd except for existing dryers or feeders to existing building w/ no other continuous metal path (see **p.4**)	[3801.4]	{338.10B2X}
☐ Bends gradual (min 5× cable diameter) _____	[3802.5]	{338.24}

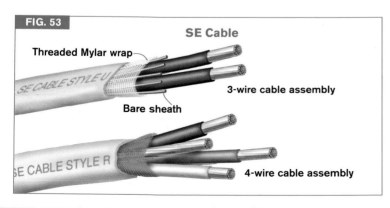

FIG. 53

SE Cable

Threaded Mylar wrap

SE CABLE STYLE U

3-wire cable assembly

Bare sheath

SE CABLE STYLE R

4-wire cable assembly

MC–Metal-Clad Cable F54,55

	09 IRC	11 NEC
☐ Support or secure at max 6ft intervals _____	[3802.1]	{330.30B&C}
☐ Secure within 12in of box or other termination EXC__	[3802.1]	{330.30B}
• Unsupported whip ≤6ft to luminaire in accessible ceiling	[3802.1]	{330.30D}
• Where fished _____	[n/a]	{330.30D}
☐ Bends gradual (min 7× interlocked armor diameter) _	[3802.5]	{330.24}

FIG. 54   **FIG. 55**

MC Cable

MC Cable Clamp (no locknut style)

Plastic wrap

Green wire

Metal armor is not a ground.

FIG. 56

Stand-Off Clamp

Used to maintain clearances to stud or joist edge

VOLTAGE DROP

When laying out wiring, consider the voltage drop caused by long runs of wire. Informational note #4 of 210.19 of the NEC recommends (though it does not require) a maximum voltage drop of 3% on branch circuits and a 5% overall voltage drop, including the feeders. Excessive voltage drop can cause problems in connected equipment and adds to the monthly utility costs. One way to overcome a voltage drop problem is to use larger wire than the minimum size and to make sure that all connections are tight. Voltage drop increases proportionately to the load on the circuit. Adding more than the minimum number of circuits helps to prevent individual circuits from overloading. The added cost of more wiring will pay for itself over time in reduced utility costs and greater equipment efficiency.

TABLE 15	CABLE LENGTH TO LIMIT VOLTAGE DROP TO 3%			
Wire Size	**Cu Distance**		**Al Distance**	
	120V	**240V**	**120V**	**240V**
14AWG	50 ft.	100 ft.	N/A	N/A
12AWG	60 ft.	120 ft.	36 ft.	72 ft.
10AWG	64 ft.	128 ft.	38 ft.	76 ft.
8AWG	76 ft.	152 ft.	46 ft.	92 ft.
6AWG	94 ft.	188 ft.	57 ft.	114 ft.

Based on 80% circuit loading for normal OCPD.

RACEWAYS

Raceways are complete systems of conduit or tubing through which conductors are installed. In the NEC numbering system, all articles pertaining to raceways have a parallel numbering system so the portion after the article number is the same for all types. Article numbers are the first 3 digits before the period inside each section number. See the common numbering system table (T24) on the inside back cover.

General Requirements for Raceways

	09 IRC	11 NEC
☐ Conductors in raceways stranded if ≥8 AWG	[3406.4]	{310.106C}
☐ Wet-rated conductors req'd in raceways above grade in wet locations	[3802.7][43]	{300.9}
☐ Raceway req to be complete prior to wiring EXC	[3904.5]	{300.18A}
• Short sections of raceway for cable protection	[3904.5X]	{300.18AX}
☐ Bends req'd to have even radius—no kinks	[3802.5]	{***.24}
☐ 360° max bends between pull points F57	[3802.1]	{***.26}
☐ Raceway must be reamed after cutting	[3802.1]	{***.28}
☐ Plastic bushing/liner req'd if conductors ≥4AWG	[3906.1.1]	{300.4G}
☐ Box & conduit body covers must remain accessible	[3905.10]	{314.29}
☐ No plastic boxes w/ metal cables or raceways unless bonded through box	[3905.3X]	{314.3X}
☐ No splicing in conduit bodies except conduit bodies w/ sufficient volume per marking	[3905.12.3.1]	{314.16C2}
☐ Max 40% fill if >2 conductors T21,22	[3904.6]	{***.22}
☐ Derate conductors as needed T11–14	[3705.2&3]	{310.15B}

FIG. 57

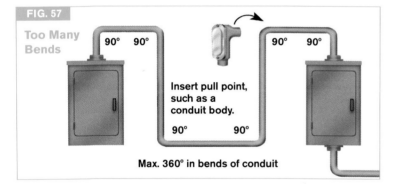

Too Many Bends

90° 90° 90° 90°

Insert pull point, such as a conduit body.

90° 90°

Max. 360° in bends of conduit

EMT–Electrical Metallic Tubing F58

	09 IRC	11 NEC
☐ Direct burial or embedment not OK	[3801.4]	{358.10B}
☐ In dry/wet locations L&L wet fittings	[3905.11]	{358.42}
☐ Secure in place max 10ft intervals & 3ft from each box, conduit body or cabinet	[3802.1]	{358.30A}
☐ Horizontal runs supported by holes in framing OK if securely fastened within 3 ft of box, conduit body, or cabinet	[3802.1]	{358.30B}
☐ Not OK as support for boxes—but OK for conduit bodies	[n/a]	{358.12}

FIG. 58

EMT–Electrical Metallic Tubing

Raintight wet location   Dry location

Sealing ring   Compression ring

Older style EMT connectors with only compression ring were not listed

ENT–Electrical Nonmetallic Tubing F64

	09 IRC	11 NEC
☐ OK embedded in concrete w/ approved fittings	[3801.4]	{362.10}
☐ Not OK in environments >50°C (122°F)	[n/a]	{362.12}
☐ Not OK for direct earth burial	[3801.4]	{362.12}
☐ Must be identified as sunlight resistant if outdoors	[3801.4]	{362.12}
☐ Secure or support every 3ft EXC	[3802.1]	{362.30A&B}
• 6ft unsupported OK to luminaires in accessible ceiling	[3802.1]	{362.30AX2}

RMC–Rigid Metal Conduit F59

	09 IRC	11 NEC
☐ Galvanized RMC typically sufficient corrosion protection for direct burial or embedment	[3801.4]	{344.10B}
☐ Coat buried field cut threads w/ L&L compound	[3801.4]	{300.6A}
☐ Provide bushing or fitting at box connection F59	[3802.1]	{344.46}
☐ No threadless connectors on threaded conduit ends	[n/a]	{344.42}
☐ Secure in place within 3ft of termination	[3802.1]	{344.30A}
☐ Horizontal support spacing max 10ft	[3802.1]	{344.30B}

FIG. 59

RMC– Rigid Metal Conduit

Box wall

Interior reamed   Locknut

Bushing
Locknut
Locknut

FMC–Flexible Metal Conduit ("Greenfield") F60

	09 IRC	11 NEC
☐ Dry locations only	[3801.4][44]	{348.12}
☐ Support max spacing 4½ft & 12in from boxes EXC	[3802.1]	{348.30A}
• Lighting whip in accessible ceiling OK to 6ft OR	[3802.1]	{348.30AX4}
• 36in where flexibility is needed	[3802.1]	{348.30AX2}
☐ Armor is OK as EGC if fittings listed, circuit ≤20A, no flexibility needed, & ≤6ft long	[3908.8.1]	{250.118}
☐ Angle connections may not be concealed F60	[n/a]	{348.42}

FIG. 60

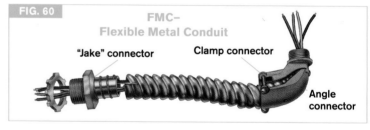

FMC– Flexible Metal Conduit

"Jake" connector   Clamp connector

Angle connector

LFMC–Liquidtight Flexible Metal Conduit F61

	09 IRC	11 NEC
☐ OK for wet locations	[3801.4]	{350.10}
☐ OK for direct burial if L&L	[3801.4]	{350.10}
☐ OK as EGC up to 6ft if fittings listed, circuit ≤20A or ≤60A for sizes ³/₄in–1¼in, & no flexibility needed	[3908.8.1]	{250.118}
☐ Support max spacing 4½ft & 12in from boxes EXC	[3802.1]	{350.30A}
• 36in where flexibility is needed	[3802.1]	{350.30AX2}

FIG. 61

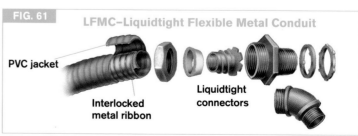

LFMC–Liquidtight Flexible Metal Conduit

PVC jacket

Liquidtight connectors

Interlocked metal ribbon

LFNC–Liquidtight Flexible Nonmetallic Conduit F62

	09 IRC	11 NEC
☐ OK in lengths >6ft if secured every 3ft	[n/a]	{356.10}
☐ Securing or supporting not req'd up to 3ft for motors	[3802.1]	{356.30}
☐ OK for direct burial or encasement when L&L	[3801.4]	{356.10}
☐ EGC req'd	[3908.4]	{250.4A5}

PVC – Rigid Polyvinyl Chloride Conduit F63   09 IRC   11 NEC

- ☐ Burial depth per T1 _____ [3803.1] {300.5A}
- ☐ Support to prevent sags per T16 & within 3ft of box _ [3802.1] {352.30}
- ☐ Expansion joints req'd if subject to ≥$\frac{1}{4}$in shrinkage_____ [n/a] {352.44}
- ☐ Not OK for support of luminaires or boxes _____ [n/a] {352.12B}
- ☐ Not permitted in environments >50°C (122°F)_____ [n/a] {352.12D}

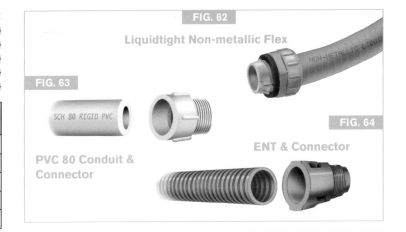

FIG. 62 Liquidtight Non-metallic Flex
FIG. 63 SCH 80 RIGID PVC — PVC 80 Conduit & Connector
FIG. 64 ENT & Connector

TABLE 16	PVC CONDUIT SUPPORT MAX. SPACING [T3802.1] {T352.30}	
Conduit Trade Size	**2009 IRC**	**2011 NEC**
$\frac{1}{2}$ in.–1 in.	3 ft.	3 ft.
1$\frac{1}{4}$ in.–2 in.	5 ft.	5 ft.
2$\frac{1}{2}$ in.–3 in.	5 ft.	6 ft.
3$\frac{1}{2}$ in.–5 in.	5 ft.	7 ft.

FILL TABLES FOR ALL CONDUCTORS OF THE SAME SIZE. BASED ON NEC CHAPTER 9 & ANNEX C

TABLE 17 — EMT FILL {Table C.1}

Size AWG kcmil	$\frac{1}{2}$	$\frac{3}{4}$	1	1$\frac{1}{4}$	1$\frac{1}{2}$	2
14	12	22	35	61	84	138
12	9	16	26	45	61	101
10	5	10	16	28	38	63
8	3	6	9	16	22	36
6	2	4	7	12	16	26
4	1	2	4	7	10	16
3	1	1	3	6	8	13
2	1	1	3	5	7	11
1	1	1	1	4	5	8
1/0	1	1	1	3	4	7
2/0	0	1	1	2	3	6
3/0	0	1	1	1	3	5
4/0	0	1	1	1	2	4
250	0	0	1	1	1	3

(Number of Conductors in THHN, THWN)

TABLE 18 — EMT FILL {Table C.1(A)}

Size AWG kcmil	$\frac{1}{2}$	$\frac{3}{4}$	1	1$\frac{1}{4}$	1$\frac{1}{2}$	2
14	–	–	–	–	–	–
12	–	–	–	–	–	–
10	–	–	–	–	–	–
8	3	5	8	15	20	34
6	1	4	6	11	15	25
4	1	3	4	8	11	18
3	1	2	3	7	9	15
2	1	1	3	6	8	13
1	1	1	2	4	6	10
1/0	1	1	1	3	5	8
2/0	1	1	1	3	4	7
3/0	0	1	1	2	3	6
4/0	0	1	1	1	3	5
250	0	1	1	1	2	4

(Number of Conductors in XHHW (Compact Stranding))

TABLE 19 — SCHEDULE 80 PVC FILL {Table C.9}

Size AWG kcmil	$\frac{1}{2}$	$\frac{3}{4}$	1	1$\frac{1}{4}$	1$\frac{1}{2}$	2
14	9	17	28	51	70	118
12	6	12	20	37	51	86
10	4	7	13	23	32	54
8	2	4	7	13	18	31
6	1	3	5	9	13	22
4	1	1	3	6	8	14
3	1	1	3	5	7	12
2	1	1	2	4	6	10
1	0	1	1	3	4	7
1/0	0	1	1	2	3	6
2/0	0	1	1	1	3	5
3/0	0	0	1	1	2	4
4/0	0	0	1	1	1	3
250	0	0	1	1	1	3

(Number of Conductors in THHN, THWN)

TABLE 20 — SCHEDULE 80 PVC FILL {Table C.9 (A)}

Size AWG kcmil	$\frac{1}{2}$	$\frac{3}{4}$	1	1$\frac{1}{4}$	1$\frac{1}{2}$	2
14	–	–	–	–	–	–
12	–	–	–	–	–	–
10	–	–	–	–	–	–
8	1	4	7	12	17	29
6	1	3	5	9	13	21
4	1	1	3	6	9	15
3	1	1	3	5	7	13
2	1	1	2	5	6	11
1	1	1	1	3	5	8
1/0	0	1	1	3	4	7
2/0	0	1	1	2	3	6
3/0	0	1	1	1	3	5
4/0	0	0	1	1	2	4
250	0	0	1	1	1	3

(Number of Conductors in XHHW (Compact Stranding))

Conduit Fill Calculations

When all conductors are the same size, use T17–20. When different sized conductors are used, use T21 to find the wire areas, add them up, and use T22 to find the minimum size conduit. Example: 3 2AWG THHN + 3 8AWG XHHW in FMC. (3 × 0.1158) + (3 × 0.0394) = 0.4656, and the next greater size in the 40% column is 0.511. Therefore, a 1$\frac{1}{4}$in. FMC conduit meets code. When conductor calculation is close to conduit table values, one size larger is recommended.

TABLE 21 — SQ. IN. AREA OF CONDUCTORS (BASED ON NEC T5 CHAPTER 9)

	14	12	10	8	6	4	2	1	1/0	2/0	3/0	4/0	250
TW	.0139	.0181	.0243	.0437	.0726	.0973	.1333	.1901	.2223	.2624	.3117	.3718	.4596
THHN	.0097	.0133	.0211	.0366	.0507	.0824	.1158	.1562	.1855	.2223	.2679	.3237	.3970
XHHW (Compact Stranding)	.0394	.0530	.0730	.1017	.1352	.1590	.1885	.2290	.2733	.3421			

TABLE 22 — CONDUIT & TUBING FILL (BASED ON NEC T4 CHAPTER 9)

Trade Size	Internal Area								2 wire sq. in. Fill 31%								>2 Wire sq. in. Fill 40%							
	EMT	ENT	FMC	LFMC	IMC	RMC	PVC80	PVC40	EMT	ENT	FMC	LFMC	IMC	RMC	PVC80	PVC40	EMT	ENT	FMC	LFMC	IMC	RMC	PVC80	PVC40
$\frac{3}{8}$	–	–	.116	.192	–	–	–	–	–	–	.036	.059	–	–	–	–	–	–	.046	.077	–	–	–	–
$\frac{1}{2}$	.304	.246	.317	.314	.342	.314	.217	.285	.094	.076	.098	.097	.106	.097	.067	.088	.122	.099	.127	.125	.137	.125	.087	.114
$\frac{3}{4}$	.533	.454	.533	.541	.586	.549	.409	.508	.165	.141	.165	.168	.182	.170	.127	.157	.213	.181	.213	.216	.235	.220	.164	.203
1	.864	.785	.817	.873	.959	.887	.688	.832	.268	.243	.253	.270	.297	.275	.213	.258	.346	.314	.327	.349	.384	.355	.275	.333
1$\frac{1}{4}$	1.496	1.410	1.277	1.528	1.647	1.526	1.237	1.453	.464	.437	.396	.474	.510	.473	.383	.450	.598	.564	.511	.611	.658	.610	.495	.581
1$\frac{1}{2}$	2.036	1.936	1.858	1.981	2.225	2.071	1.711	1.986	.631	.600	.576	.614	.689	1.056	.530	.616	.814	.774	.743	.792	.889	.829	.684	.794
2	3.356	3.205	3.269	3.246	3.630	3.408	2.874	3.291	1.040	.994	1.013	1.006	1.125	1.508	.891	1.020	1.342	1.282	1.307	1.298	1.452	1.363	1.150	1.316

SWIMMING POOL

Electricity and water can be a lethal mix. Precautions must be taken for shock hazard protection and to prevent corrosion of electrical equipment. Bonding is important to eliminate voltage gradients in the pool area. For GFCI requirements, see **p.16**. Installation of a pool might require relocating overhead service conductors.

Overhead Conductor Clearances	09 IRC	11 NEC
☐ 22½ft clearance in any direction from water	[T4203.5]	{680.8A}
☐ 14½ft in any direction from diving platform	[T4203.5]	{680.8A}

Underground Wiring	09 IRC	11 NEC
☐ Non-pool underground wiring min 5ft from pool EXC	[4203.7]	{680.10}
• If space limited, RMC, IMC, or PVC systems OK	[4203.7]	{680.10}
☐ Cover depth min 6in for RMC or IMC, 18in for PVC	[T4203.7]	{680.10}

Feeders to Pool Panelboards	09 IRC	11 NEC
☐ New feeder req's RMC, IMC, LFNMC, or PVC EXC	[T4202.1]	{680.25A}
• EMT OK on or within buildings	[T4202.1]	{680.25A}
☐ Raceway req's min 12AWG insulated EGC EXC	[T4202.1]	{680.25B}
• Existing FMC or cable w/ EGC OK	[T4205.6]	{680.25AX}

Pool Pump Motors	09 IRC	11 NEC
☐ RMC, IMC, PVC, or listed MC OK for branch circuit	[T4202.1]	{680.21A1}
☐ Branch circuits in AC, FMC, NM only within building	[T4202.1]	{680.21A1}
☐ EMT branch circuit OK on or within building	[T4202.1]	{680.21A2}
☐ Flexible connection OK in LFMC or LFNMC	[T4202.1]	{680.21A3}
☐ Cord & plug connected motors OK w/ cord ≤3ft	[4202.2]	{680.21A5}
☐ Cords req EGC min 12AWG & per **T6**	[4202.2]	{680.7B}

Underwater Wet-Niche Lighting F68	09 IRC	11 NEC
☐ Min 18in below water level	[4206.4.2]	{680.23A5}
☐ Luminaire bonded & secured to shell w/ locking device	[4206.5]	{680.23B5}
☐ Luminaire must req tool for removal	[4206.5]	{680.23B5}
☐ Low-voltage transformers req L&L for pool	[4206.1]	{680.23A2}
☐ Conductors from load side of GFCI or transformer not in same raceway or box as non-GFCI wires	[4206.3]	{680.23F3}
☐ Forming shell req's bonding terminal if PVC conduit	[4206.5]	[680.23B1]
☐ Nonmetallic conduit req's 8AWG bonding conductor	[4205.3]	{680.23B2}
☐ Bonding conductor insulated & potted in forming shell	[4205.3]	{680.23B2}
☐ Min 16AWG EGC in cord to wet-niche fixture	[4205.4]	{680.23B3}
☐ EGC connections on terminals only—no splices	[4205.2]	{680.23F2}

Equipotential Bonding F68	09 IRC	11 NEC
☐ Purpose of bonding is to reduce voltage gradients	[4204.1]	{680.26A}
☐ Bond metal parts of pool structure, ladders, equipment, fences, & screens or structures <5ft from pool EXC	[4204.2]	{680.26B}[47]
• Small isolated parts <4in or <1in into pool structure	[4204.2]	{680.26B5}
☐ Bond motors except listed & double-insulated type	[4204.2]	{680.26B6X}
☐ Provide bond wire to area of double-insulated motor	[4204.2]	{680.26B6}
☐ Bonding conductor min #8 solid Cu	[4204.4]	{680.26B}
☐ Unencapsulated steel shell req'd to be bonded	[4204.2]	{680.26B1}
☐ Cu conductor grid req'd if pool shell steel encapsulated in non-conductive compounds (coated rebar)	[4204.2][48]	{680.26B1}
☐ Cu conductor grid req's 8AWG Cu in 12in ×12in pattern, conforming to contour of pool & deck, ≤6in from outer contour of pool shell, all conductors bonded at crossings	[4204.2][48]	{680.26B1}
☐ Perimeter surfaces for 3ft beyond pool req equipotential bonding w/ steel wire or reinforcement	[4204.2][48]	{680.26B2b}
☐ Connect perimeter to unencapsulated steel pool shell or Cu conductor grid at min 4 points	[4204.2][48]	{680.26B2}
☐ Min 9sq in bonded metal in contact w/ pool water	[4204.3][49]	{680.26C}

Receptacles (see p.16 for GFCI requirements)	09 IRC	11 NEC
☐ Min 1 receptacle from pool walls	[4203.1.2]	{680.22A3}
☐ Pump motor receptacles not <10ft from pool wall EXC		
• 6ft OK for single-receptacle twist-lock types	[4203.1.1]	{680.22A1}
☐ Dimensions include distance around barriers w/o penetrating a floor, wall, doorway, or window opening	[4203.1]	{680.22A5}

Lighting Outlets & Luminaires	09 IRC	11 NEC
☐ Outdoors min 5ft from pool edge unless 12ft above	[4203.4.1]	{680.22B1}
☐ Indoors ≥7ft 6in above water if enclosed & GFCI	[4203.4.2]	{680.22B2}
☐ Existing OK if GFCI & ≥5ft from pool edge & ≥5ft high	[4203.4.3]	{680.22B3}
☐ Switches min 5ft from pool edge or separated by barrier	[4203.2]	{680.22C}

HOT TUB/SPA

Outdoor hot tubs and spas follow the same rules as swimming pools in addition to the general rules below. A hydromassage tub (**p. 20**) is not a spa because it is emptied after each use.

General	09 IRC	11 NEC
☐ LFMC or LFNMC up to 6ft OK for package unit	[T4202.1]	{680.42A1}
☐ Cord up to 15ft OK for GFCI-protected package unit	[4202.2]	{680.42A2}
☐ Bands to secure hot tub staves exempt from bonding	[4204.4]	{680.42B}

Indoor Spas	09 IRC	11 NEC
☐ Indoor packaged units ≤20A OK for cord & plug	[4202.2]	{680.43X}
☐ Min 1 receptacle 6ft–10ft from inside wall of spa	[4203.1.4]	{680.43A1}
☐ Wall switches min 5ft from inside wall of spa	[4203.2]	{680.43C}

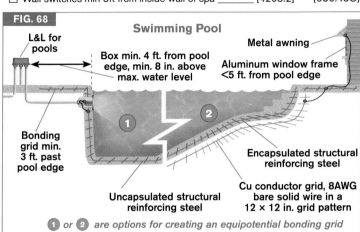

FIG. 68

Swimming Pool

L&L for pools

Box min. 4 ft. from pool edge, min. 8 in. above max. water level

Metal awning

Aluminum window frame <5 ft. from pool edge

Bonding grid min. 3 ft. past pool edge

Uncapsulated structural reinforcing steel

Encapsulated structural reinforcing steel

Cu conductor grid, 8AWG bare solid wire in a 12 × 12 in. grid pattern

①1 or ②2 are options for creating an equipotential bonding grid

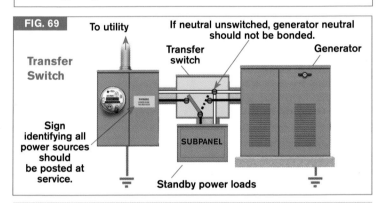

FIG. 69

To utility

If neutral unswitched, generator neutral should not be bonded.

Transfer Switch

Transfer switch

Generator

Sign identifying all power sources should be posted at service.

SUBPANEL

Standby power loads

GENERATORS

Generators provide a source of emergency power during a utility outage. Care must be taken to ensure that the two sources of power–utility and generator–cannot be connected simultaneously. This dangerous condition results from failure to install proper transfer switches and improper use of portable generators.

Generators	11 NEC
☐ Must be suitable for environment, rainproof if outdoors	{445.10}
☐ Rainproof generators not OK enclosed indoors	{110.3B}
☐ Conductors sized 115% of nameplate current rating	{445.13}
☐ Live or moving parts guarded against accidental contact	{445.14}
☐ GEC req'd for permanently installed generators	{250.30A5}
☐ Remove bonding jumper if transfer switch does not switch neutral F69	{250.24A5}

Transfer Switches F69	11 NEC
☐ Sign req'd at service indicating generator loc	{702.7A}
☐ Transfer equip must prevent simultaneous connection of generator & utility service	{702.5}

Electric Vehicle (EV) Charging Systems	11 NEC
☐ Systems >20A 125V no exposed live parts	[625.13]
☐ Coupler L&L for EV	{625.16}
☐ Interlock must de-energize connector when uncoupled from EV	{625.18}
☐ Electric vehicle OK as standby power source through listed utility interactive connection	{625.26}

OLD WIRING

A high percentage of residential electrical fires occur in older homes. Proper overcurrent protection helps prevent insulation failure, though in some cases time and exposure take too great a toll on wiring, and it must be replaced with new materials. Fuses provide overcurrent protection only if they are the right size. Too often, they are altered or bypassed (a penny behind the fuse). Older ceramic fuse panels and panels with cartridge fuses also pose a risk of electrocution because of exposed electrical contacts. For these reasons, many insurance companies require upgrading of fuse systems. The references below are from the NEC. The IRC is a code for new construction and does not address old wiring.

Fuses   11 NEC

- ☐ No exposed contact fuseholders (must be dead front) F73 _____ {240.50D}
- ☐ Edison base (plug fuses) not OK for 240V circuits _____ {240.51A}
- ☐ Type S fuse req'd if tampering or overfusing exists F73 _____ {240.51B}
- ☐ Type S fuse adapter must be proper size for wire_____ {240.4D}
- ☐ No fuses in neutral conductor F73_____ {240.22}

KNOB & TUBE

K&T wiring is the oldest wiring method found in American homes. When left in its original state, it can be reliable; safety was inherent in its design. As a wiring method in uninsulated joist and stud cavities it is protected from damage and provided with air circulation, which prevents heat buildup. Unfortunately, when these systems are modified by unqualified persons, the inherent safety of K&T is often compromised. Adding new loads to an old system is tricky and seldom done correctly. Rubber insulation on K&T wiring becomes brittle over time and is prone to mechanical damage, especially when thermal insulation is added to an attic. Older rubber insulation has only a 60°C rating.

Knob & Tube (K&T)   11 NEC

- ☐ No new K&T _____ {394.10}
- ☐ Additions to existing K&T OK if properly protected _____ {394.10}
- ☐ Splices to other wiring methods must be in box EXC F72 _____ {300.16A}
  - • Bushing OK at termination of raceway to open switchboards ___ {300.16B}
- ☐ All conductors of circuit must enter metal box through same hole _ {300.20B}
- ☐ Must enter plastic boxes through separate holes _____ {314.17C}
- ☐ Must be protected w/ loom where entering box _____{314.17B&C}
- ☐ Loom must extend from last insulator to ¼in inside box F72 ____{314.17B&C}
- ☐ Do not envelop w/ thermal insulation _____ {394.12}
- ☐ Wires must be kept out of direct contact w/ wood framing_____{394.17}
- ☐ Tubes req'd where passing through framing members F70_____{394.17}
- ☐ 3in min between wires, 1in to surfaces F71 _____ {394.19A1}
- ☐ Conductors on sides (not face) of exposed joists & rafters EXC {394.23A&B}
  - • OK on edges or faces of rafters or joists in attics <3ft high ___ {394.23BX}
- ☐ Protect w/ running boards up to 7ft high in attic w/ stairs_____ {394.23A}
- ☐ Provide protection where exposed <7ft above floor_____ {398.15C}

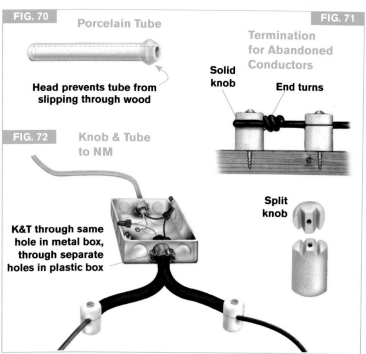

FIG. 70

Porcelain Tube

Head prevents tube from slipping through wood

FIG. 71

Termination for Abandoned Conductors

Solid knob

End turns

FIG. 72

Knob & Tube to NM

K&T through same hole in metal box, through separate holes in plastic box

Split knob

OLD NM

Pre-1984 nonmetallic sheathed cable contained conductors with insulation rated 60°C. When installed in a hot attic, the ampacity of this old wire is easily exceeded. Precautions must also be taken to isolate this old low-temperature wiring from luminaires that require high-temperature rated connections F36. Much of this old wire was used in houses with problematic electrical equipment. Replacement circuit breakers for older panels can be very expensive–providing one more incentive to replace such systems. For further information on old wiring, refer to the Code Check website, www.codecheck.com.

Aluminum Wiring   11 NEC

- ☐ Snap switches w/ direct Al connection req L&L as "CO/ALR" ___ {404.14C}
- ☐ Receptacles ≤20A w/ direct Al connection req L&L as "CO/ALR" _ {406.3C}
- ☐ Al to Cu splicing devices must be listed for same _____ {110.14}
- ☐ Terminals (including breaker terminals) for Al req L&L _____ {110.14A}

Pre-1984 NM   11 NEC

- ☐ Derate for ambient temp _____ {310.15B2}
- ☐ No 60°C conductors in attics >131°F_____ {T310.15B2}
- ☐ No direct connection to luminaires that req >60°C conductors __ {410.117B}
- ☐ Isolate old wiring from high-temp wiring F36 _____ {410.117A}
- ☐ Box for tap conductors min 1ft from luminaire, max 6ft wire _____ {410.117C}

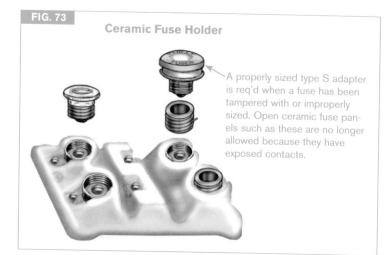

FIG. 73

Ceramic Fuse Holder

A properly sized type S adapter is req'd when a fuse has been tampered with or improperly sized. Open ceramic fuse panels such as these are no longer allowed because they have exposed contacts.

REPLACEMENT RECEPTACLES

Houses built before adoption of the 1962 NEC will not have 3-hole receptacles in all locations. Appliances with 3-prong cords are designed to be used only with grounded 3-hole receptacles. A GFCI can provide shock hazard protection for two-conductor circuits; though without an EGC, it may not protect equipment.

General   11 NEC

- ☐ AFCI protection req'd for replacements in areas where circuit req's AFCI protection (p. 13) effective 1/1/2014_____ {406.4D4}[50]
- ☐ Protection can be breaker, AFCI outlet device, or upstream AFCI outlet _____ {406.4D4}[50]
- ☐ Replacement receptacles must be tamper-resistant _____ {406.4D5}[50]
- ☐ Outdoor wet location replacement receptacles must be WR ___ {406.4D6}[50]

Replacements When No Grounding Present   11 NEC

- ☐ 2-hole receptacle OK if in area where GFCI not req'd _____ {406.4D2a}
- ☐ Must have GFCI protection in area that now req's GFCI _____ {406.4D3}
- ☐ OK to install GFCI even if no ground present_____ {406.4D2b&c}
- ☐ Non-grounded GFCI or GFCI-protected receptacles req label stating "No Equipment Ground" _____ {406.4D2b}
- ☐ Ungrounded 3-hole receptacle supplied through a GFCI also req label stating "GFCI Protected" _____ {406.4D2d}
- ☐ Separate EGC can be added from receptacle box & connect to service enclosure, GEC, or ground bar of panel at circuit origin _ {250.130C}
- ☐ OK to run EGC separately from circuit conductors _____ {300.3B2}
- ☐ Not OK to jumper neutral & EGC_____ {250.142B}

Replacements When Grounding Present in Box   11 NEC

- ☐ Replacements must be 3-hole if EGC present _____ {406.4D1}
- ☐ Bond 3-hole receptacle to grounded box w/ wire OR _____ {250.146}
  - • Use grounding-type receptacle (captive metal screw from yoke) {250.146B}